Circumcision and Human Rights

George C. Denniston • Frederick Mansfield Hodges
Marilyn Fayre Milos
Editors

Circumcision and Human Rights

Springer

Editors
George C. Denniston
University of Washington
Seattle, Washington
USA

Marilyn Fayre Milos
National Organization of Circumcision
Information Resource Centers
San Anselmo, California
USA

Frederick Mansfield Hodges
Berkeley, California
USA

ISBN 978-1-4020-9166-7 e-ISBN 978-1-4020-9167-4

Library of Congress Control Number: 2008935888

Printed on acid-free paper

springer.com

Preface

This volume is one of a series of publications based on symposia sponsored by the National Organization of Circumcision Information Resource Centers (NOCIRC), all asserting that infant and child circumcision are harmful, anachronistic practices deserving to be relegated to the history of medical error. The authors, representing a broad spectrum of personal and professional backgrounds, hope that their diverse contributions will be applied to answering a single question: Aside from extremely rare medical emergencies, is there ever any justification for removing a normal and functionally essential part of an infant or child's genital anatomy?

Most parents who agree to circumcision for their newborn sons, and most physicians who perform circumcisions, do so because they believe that it is the right thing to do. Knowing virtually nothing about the vital functions of the foreskin, they think of circumcision as somehow more "hygienic" or "attractive" than intact genitals. We argue that neither of those is the case. Although the United States has an extraordinarily high infant circumcision rate — presently about 56%,[1] far higher than that of any other Western nation — few people ever ask why we have preserved this outdated practice. Parents accept it because others do so, because older family members expect it, or because a physician or nurse hands them a consent form to be signed. Perhaps asking the simplest of questions — Why? — seems improper or inappropriate when the subject is genitals. But the authors of this volume think this is precisely the question that must always be posed. Here we present some of our reasons for answering as we do.

Even among those religious groups that promote ritual circumcision of infant sons, we find that many parents do so only because they believe that they must uphold tradition and that the procedure confirms their allegiance to the religious community. Desire for community cohesion and respect for a past way of life overcome the natural inclination of the modern mind to reject genital cutting of an infant. Still, respect for a tradition, no matter how ancient or venerated, should not be used to violate another person's right to physical integrity.[2] We believe that, aside from the demands of obvious medical emergencies, everyone has the right to keep all of his or her natural body parts.[3]

In the late nineteenth century, when calls for circumcision first appeared in the American medical community, claims for the virtues of the practice would now seem laughable: it would prevent boys from masturbating; it was a potential cure for epilepsy, for various orthopedic disorders, even for insanity.[4] But as the nation steadily modernized, and as more people achieved better understanding, not only of human anatomy and physiology but of social motivation and religious psychology, new kinds of rationalization for circumcision appeared — but phrased in the language of modern medicine. In the twentieth century, it was no longer masturbation or insanity that would be "cured" by foreskin removal, but a succession of more contemporary fears. Predictably, it was the scourge of the century that attracted most attention, and from one decade to another circumcision was credited with prevention of penile cancer,[5] prostatic cancer,[6] and even cervical cancer.[7] (This last claim, an instructive case study in medical history, was based on observations that Jewish women seemed less susceptible to the disease — but without consideration of the many other equally plausible explanations. The original author of the claim eventually retracted it.[8])

Over the years, each new claim for the benefits of circumcision has been painstakingly refuted, although with lagging effect on public opinion. Most recently, the public has been told — with overwhelming hyperbole, inappropriate to scientific inquiry[9,10] — that circumcision will prevent the newest scourge, HIV infection and its ultimate result, the potentially fatal immune deficiency syndrome called AIDS. This latest claim is based primarily on three parallel, coordinated randomized controlled trials[11–13] conducted among men in eastern and southern Africa — the only part of the world where the HIV/AIDS epidemic has become a major problem among heterosexuals. Although those studies were initially accepted with enthusiasm,[14] as the air cleared there began to appear a number of publications raising questions about methodology, long-term benefits for individuals, and actual results for affected populations.[15–17]

But conclusions about those studies, one way or another, cannot and should not be cited to justify continuation of infant circumcision in America. As everyone in the public health community is well aware, the HIV/AIDS problem in this country (and indeed worldwide) differs strikingly from that in eastern and southern Africa. The disease in this country has been confined largely to homosexual men, drug addicts, and communities where sexual promiscuity, inadequate information, and indifference to prevention are a serious problem.[18,19] The vast majority of middle-class heterosexual Americans do not live in fear of HIV infection, and there is no reason to anticipate that their newborn infants will face an epidemic 20 or 30 years hence. Evidence for declining fear of AIDS is found throughout the world: Europe, Latin America, nearly all of Asia — where circumcision rates are very low or even zero — have impressively low infection rates, in most cases significantly lower than our own. In short, studies of adults in a distant continent, where sexual culture and sexual practices differ radically from our own, have no legitimacy with regard to infant circumcision in America or elsewhere.

This symposium volume, like its predecessors, addresses a variety of issues connected with circumcision worldwide. We hope that it will contribute to the public's steadily growing awareness that excising parts of infant and child genitals should have no place in medical practice.

George C. Denniston
Frederick M. Hodges
Marilyn Fayre Milos

References

1. National Hospital Discharge Survey for 2006. National Center for Health Statistics, Hyattsville, MD, 2008.
2. Article 18(3), United Nations Covenant on Civil and Political Rights. United Nations General Assembly Resolution 2200A [XX1], 16 December 1966.
3. UNESCO. Universal Declaration on Bioethics and Human Rights. Adopted by the UNESCO General Conference, 19 October 2005.
4. Gollaher DL. From ritual to science: the medical transformation of circumcision in America. J Soc Hist 1994;28(1):5–36.
5. Wolbarst AL. Circumcision and penile cancer. Lancet 1932;1(5655):150–3.
6. Apt A. Circumcision and prostatic cancer. Acta Med Scand 1965; 178:493–504.
7. Wynder E. Circumcision as a preventive factor against cancer of the cervix. Proceedings of the third national cancer conference 603–7, 1956.
8. Wynder EL, Licklider SD. The question of circumcision. Cancer 1960;13(3):442–5.
9. Van Howe RS, Svoboda JS, Hodges FM. HIV infection and circumcision: cutting through the hyperbole. J R Soc Health 2005;125(6):259–65.
10. Garenne M. Male circumcision and HIV control in Africa. PLoS Med 2006;3(1):e78.
11. Auvert B, Taljaard D, Lagarde E, et al. Randomized, controlled intervention trial of male circumcision for reduction of HIV infection risk: The ANRS 1265 trial. PLoS Med 2005;2:e298.
12. Bailey RC, Moses S, Parker CB, et al. Male circumcision for HIV prevention in young men in Kisumu, Kenya: a randomised controlled trial. Lancet 2007;369(9562), 643–56.
13. Gray RH, Kigozi G, Serwadda D, et al. Male circumcision for HIV prevention in men in Rakai, Uganda: a randomised trial. Lancet 2007;369(9562), 657–66.
14. Sawires SR, Dworkin SL, Fiamma A, et al. Male circumcision and HIV/AIDS: challenges and opportunities. Lancet 2007;369:708–13.
15. Dowsett GW, Couch M. Male circumcision and HIV prevention: is there really enough of the right kind of evidence? Reprod Health Matters 2007;15(29):33–44.
16. Green LW, McAllister RG, Peterson KW, Travis JW. Male circumcision is not the HIV 'vaccine' we have been waiting for. Future HIV Ther 2008;2(3):193–9.
17. Garenne M. Long-term population effect of male circumcision in generalised HIV epidemics in sub-Saharan Africa. Afr J AIDS Res 2008;7(1):1–8.
18. Anonymous. CDC HIV/AIDS Science Facts: Male Circumcision and Risk for HIV Transmission and Other Health Conditions: Implications for the United States. Atlanta: Centers for Disease Control, 2008.
19. Millett GA, Ding H, Lauby J, et al. Circumcision status and HIV infection among Black and Latino men who have sex with men in 3 US cities. J Acquir Immune Defic Syndr 2007;15;46(5):643–50.

Acknowledgments

Many dedicated people were instrumental in the production of this volume and the symposium out of which it grew. We would like to offer our most heartfelt thanks to Ken Brierley, Sheila Curran, and Chelo Jacob for running the symposium with efficiency and grace. John Geisheker, JD, Executive Director, along with the staff and board of directors of Doctors Opposing Circumcision (D.O.C.), who co-sponsored the symposium, are especially worthy of commendation. We would also like to extend our gratitude to Jess Grant, Betty Katz Sperlich, and Ron Romanovsky for their part in making the symposium enjoyable as well as thought provoking. Additionally, it is a great pleasure to acknowledge all those working to defend the genital integrity of infants and children. Finally, we wish to applaud parents who have the courage to honor, respect, and protect the rights of their sons and daughters to bodily integrity and the full compass of human dignity to which every human being is entitled.

Contents

Contributors

Zachary Androus, BA, MA, and PhD Candidate in Race, Gender and Social Justice, Department of Anthropology, American University. His research interests include neonatal circumcision in the United States and the relationship between transgressive body modifications and sexuality. He teaches anthropology at the Lorenzo De Medici University. Florence, Italy.

Suzanne Arms is an author, photojournalist, film maker, founding member of the Coalition to Improve Maternity Services (CMS), founding board member of the Alliance for Transforming the Lives of Children (aTLC), and founder and director of Birth the Future. Bayfield, CO, USA.

David Chamberlain, PhD, is a California psychologist and one of the international pioneers creating the emerging field of prenatal and birth psychology. In a landmark study with mothers and child pairs in 1980, he proved that birth memories were reliable memories and not fantasies. His book, The Mind of Your Newborn Baby (1998) is in its 3rd edition, making its way around the world in twelve languages.

In 1983, David helped to create the Association for Prenatal and Perinatal Psychology and Health (APPPAH), served as its President for eight years, and has served on its board of directors ever since. He is also Editor of Birthpsychology.com, APPPAH's popular website, which he founded in 1996.

Dr. Chamberlain has taken his message about intelligent prenates and newborns to sixteen countries and his fifty publications are studied and quoted around the world. Nevada City, CA, USA.

Georganne Chapin is President and CEO of Hudson Health Plan, a not-for-profit Medicaid managed-care company in New York's Hudson Valley. She is also an adjunct faculty member of Pace University School of Law, where she has taught courses in Medicaid and disability law, as well as Bioethics. Georganne holds an undergraduate degree in anthropology, a Masters in Sociomedical Sciences, and a JD, and she is on the Board of Directors of Attorneys for the Rights of the Child (ARC). Tarrytown, NY, USA.

John Dalton, MS, received his first degree in biochemistry and his Master's Degree in Radiation Biophysics. Since 1984, he has been researching the medical literature in relation to the foreskin and its removal and has accumulated an archive of more than 3,000 documents related to this subject. He has concentrated on accumulating literature showing medical necessity for therapeutic male circumcision is doubtful.

Marie Fox is Professor of Law at the University of Keele. Her main research interests are in the fields of Health Care Law, Animal Law, and Feminist Legal Theory. Selected recent publications include: (with Jean McHale), 2nd edition of Health Care Law: Text, Cases and Materials (Sweet & Maxwell) 2006 (1204, xxxvi pages); "The Regulation of Xenotransplantation in the United Kingdom After UKXIRA: Legal and Ethical Issues" (with L.Williamson and S. McLean) (2007) 34(4) Journal of Law & Society 441–64; "Rethinking the Animal/Human Boundary: the impact of xeno technologies" (2005) 26 Liverpool Law Review 149–67; (with Michael Thomson) "Cutting it: surgical interventions and the sexing of children" (2005) 12 Cardozo Journal of Law & Gender 82–97; (with Michael Thomson) "A Covenant with the Status Quo? Male Circumcision and the new BMA Guidance to Doctors," (2005) 31 Journal of Medical Ethics 463–9; (with Michael Thomson) "Short Changed? The Law and Ethics of Male Circumcision," (2005) 13 International Journal of Children's Rights 161–81; republished in M. Freeman (ed.) Children's Health and Children's Rights Leiden/Boston: Martinus Nijhoff Publishers, 2006. Staffordshire, UK.

Noel Gazzano, PhD, is a medical anthropologist who, after working as staff anthropologist at the Centro Ricerche in Ergonomia, Firenze, received her doctoral degree from the University of Firenze, with the dissertation, Biomedical Rituals: Woman and the Embodiment of Gender in Italian Healthcare Environments. Her current research focuses on male and female body modifications in contemporary and historical medical environments, ranging from aesthetic surgery to FGM and circumcision. Florence, Italy.

Leonard Glick MD (University of Maryland), PhD (University of Pennsylvania), Professor Emeritus of Anthropology, Hampshire College, Amherst, Massachusetts (retired), is the author of Abraham's Heirs: Jews and Christians in Medieval Europe "Religion and Genocide," in I.W. Chaney (ed.), The Widening Circle of Genocide (1994), articles on religion and ethnicity, and author of Marked in Your Flesh: Circumcision from Ancient Judea to Modern America (Oxford University Press, 2005). New Salem, MA, USA.

Pia Grassivaro Gallo, PhD, Associate Professor of Anthropology, University of Padua's Psychology Faculty, and former teacher of Applied Biology, Human Genetics, and Anthropogenetics. Her research on the biology of current human populations has taken place in several developing countries, particularly Somalia (from 1972 to 1985). At the invitation of the Somali Ministry of Public Health (1981), she was invited to take part in a scientific mission to Somaliland. From 1988, she has been responsible for the Padua Working Group on FGM, dealing with African immigrants in Italy. From 2000, she studied the expansive forms of the traditional interventions on female genitalia, carrying out field research in Central Africa (Uganda, Malawi, and Congo RDC). She was co-

coordinator of the VIII International Symposium on Circumcision and Human Rights. Padua, Italy.

Matthew Hess is the founder and President of MGMbill.org, a San Diego-based group seeking to pass US laws that would protect boys from non-therapeutic circumcision. San Diego, CA, USA.

George Hill is the Vice President for Bioethics and Medical Science, Doctors Opposing Circumcision. Seattle, WA, USA.

Yngve Hofvander, MD, is a pediatrician and Professor, International Child Health, Uppsala University, consultant to the Swedish International Development Authority, World Health Organization, and UNICEF. Uppsala, Sweden.

Primus Lake, PhD, is a Professor at Nusa Cendana University in West Timor, Indonesia. As an Indonesian researcher in Southeast Asia, he researched genital enhancement practices, which was funded by the Ford Foundation. In 1988, he established Bina Insan Mandiri (Yabim) Foundation based in Kupang, Timor, to promote healthy circumcision and sexual health. Kupang, West Timor, Indonesia.

Romeo Lee, PhD, is Associate Professor, Behavioural Sciences, De La Salle University-Manila. His area of specialization is Filipino men's health, a subject he has pursued through research since the 1990s. He is the Immediate Past President and a current member of the Board of Trustees, Philippine Population Association, and is an Editorial Board Member of the International Journal of Men's Health. Manila, Philippines.

David Llewellyn, JD, a trial lawyer with the firm of Johnson & Ward in Atlanta, Georgia, has more than ten years' experience litigating wrongful circumcision, circumcision damage, and related cases in federal and state courts throughout the United States and Canada about the problem of neonatal male circumcision. A majority of his practice consists of genital injury litigation. Atlanta, GA, USA.

Ken McGrath, VRD, Msc(Hons), LIBiol, MNZIML, Senior Lecturer in Pathology in the Faculty of Health, Auckland, University of Technology, New Zealand, has made a lifelong study of the male genitalia, which he has taught to medical students. His research interests are the innervation of the penis and fungal diseases of the skin. Auckland, NZ.

Debra Moro is a graduate in Psychology and a member of the Working Group on FGM, University of Padua. Padua, Italy.

Miriam Pollack, an educator in private practice and the Jewish mother of two circumcised sons, has been advocating for genital integrity for Jewish as well as non-Jewish baby boys for the past 15 years by writing, speaking, counseling Jewish parents, and providing alternative brit b'lee milah ceremonies for interested parents. Boulder, CO, USA.

David Smith is the general manager of NORM-UK and editor of NORM NEWS. Stone, Staffordshire, UK.

James L. Snyder, MD, FACS, is a Diplomate of the American Board of Urology, a Fellow of the American College of Surgeons, Past President of the Virginia Urological Society, Retired Commander, Medical Corps, United States Naval Reserve, who retired from active medical practice in 2000. He has served as expert witness in several circumcision lawsuits. Clifton Forge, VA, USA.

Mana Sultan Abdurahman Ali Issah, Water for Life-Merka (NGO). Somalia, Africa.

J. Steven Svoboda, JD, focuses on civil litigation and human rights, and is the founder and Executive Director of Attorneys for the Rights of the Child (ARC), a non-profit organization addressing the illegality of involuntary genital surgery. Berkeley, CA, USA.

Michael Thomson is Professor of Law, Culture & Society at the University of Keele. His research interests include Health Care Law, Law and Gender, and Law and Literature. His particular focus has been the regulation of reproduction and the relationship between law and gender. The focus of his most recent work is masculinity and the legal regulation of the male sexed body. He is the author of Reproducing Narrative: Gender, Reproduction and Law (Dartmouth, 1998) and Endowed: Regulating the Male Sexed Body (Routledge, 2007). Staffordshire, UK.

Franco Viviani, PhD, a physical anthropologist, is Professor of Functional Anthropology at the Faculty of Medicine, University of Udine, and Professor of Psychobiology and Physiology of Human Behaviour at the Faculty of Psychology, University of Padova. He has published papers and articles both on male and female circumcision. He is the representative for NOCIRC of Italy. University of Padua, Italy.

Elena Vladimirova graduated as an actor from the College of Arts in Russia (1980). In 1981, she began working closely with waterbirth pioneer Igor Charkovsky. She organized birth camps at the Black Sea, delivering babies with wild dolphins participating. Together, Vladimirova and Charkovsky launched a program of natural delivery and birth trauma release for adults as well as for babies.

In 2006, Elena produced and directed a revolutionary documentary, Birth As We Know It, about the correlation between birth trauma and the quality of our life, and about what it really takes to prepare for conscious birth and to deliver a baby gracefully, minimizing trauma for all involved. She juxtaposed natural, gentle childbirth with harsh hospital practices, for example, Cesarean section, immediately cutting the umbilical cord, and circumcision, exposing the importance of gentle birthing choices. The documentary, now in 49 countries, has been translated into ten languages. Elena travels internationally, teaching her "Birth Into Being" workshops and speaking. See www.birthasweknowit.com. Chico, CA, USA.

Hugh Young, BSc, graduated in zoology from the University of Cantebury, worked as a journalist and radio producer, and, from 1994 to 1995, reviewed the literature on safe sex for the New Zealand AIDS Foundation. He created and maintains a human rights website, www.circumstitions.com, a resource of references, quotations, arguments, and visuals, particularly for intact males and their parents, treating intactness rather than circumcision as normal. He has an interest in linguistics and lexicography (as the website's glossary shows) and coined the word "circumstition." Pukerua Bay, New Zealand.

Delusional Psychologies of Circumcision and Civilization

David B. Chamberlain

Abstract The author brings 48 years of clinical experience — half of it as a pioneer in birth psychology — to a critical analysis of the delusions of pediatricians, parents, and tribal practitioners of male and female genital cutting and the similar delusions of politicians, professionals, and parents about creating a "civilization" built on violence. In the end, the author looks to a critical mass of independent and humane parents who create loving families as the only real basis for any future civilization worthy of the name.

Introduction

In one way or another, all of us are in the baby business. Of course, the whole human species is in the baby business, but that doesn't mean we understand babies; we don't. Although we were all babies once, that doesn't mean we understand who we were as babies; we really don't. Nor are we automatically aware of how much we were marked and traumatized at birth and after birth, including — God forbid — some form of genital mutilation. Getting clear about this may be delayed for decades.

The babies we care about are amazing and mysterious beings, and, as we are slowly discovering, the mind of a baby is the most mysterious part. As adults — steeped in the pop scientific culture of the twentieth century — we have had to grope our way through a matrix of myths and delusions in psychology and medicine to come anywhere near the truth. In this paper, I will name some of those delusional ideas in the hope of liberating ourselves more fully from them.

Finding the Real Baby

In virtually all cultures East and West for the last several generations, most of us have suffered from bad ideas about babies, not because of ancient myths and "old wives tales" but because of a new catechism of wrong beliefs spawned (alas!) by

---

Founding faculty and Lecturer, Santa Barbara Graduate Institute

G. Denniston et al. (eds.), *Circumcision and Human Rights*,

professionals dealing with our babies. As a parent and a psychologist, I was part of that scene. In fact, before I knew anything about the mind of a baby, my wife and I had two beautiful boys and followed the advice of our obstetrician to circumcise them. We were totally naïve in believing it was a beneficial procedure and that babies would have no memory of the experience. In retrospect, there were two sets of delusions here: the doctor's and ours. Fortunately, we had read about Grantly Dick-Read and natural childbirth[1] and did our best to approximate it but, in the 1950s, no one warned us about circumcision.

My real education about the mind of a newborn baby came suddenly in 1974 (a full 16 years after graduating with a PhD in psychology), when I took a course in the clinical applications of hypnosis and encountered lucid memories of traumas — including birth and circumcision. To put it briefly, I discovered that babies, no matter how immature they were supposed to be, always cared about what was happening to them, were learning who to trust or distrust, and were trying to put meaning to their painful experiences. They were protesting desperately how they were being treated at birth, and were trying to warn doctors about the danger of obstetrical maneuvers being used on them! Eventually, I came to understand they were developing good and bad ideas about themselves, their parents, their doctors, and life.

The newborns I encountered in hypnosis had all the senses and emotions you could ask for; they reacted fiercely to violence, usually loved their mothers, and sometimes even knew telepathically what their mothers were thinking and feeling in other parts of the hospital. Their faculties of memory and learning were obviously working but not always the way we would want. They were being hurt, traumatized, and shocked. I found I had to invent new therapeutic methods to help them resolve the host of psychological wounds and insults that were left from their encounters with parents and with modern birth.

It was only a matter of time until I realized that physicians and psychologists were teaching the world about a completely different baby with a primitive brain incapable of accurate perception, memory, or learning, with neither senses nor emotions, and not yet equipped with receptors for pain. I learned that doctors and nurses did many painful things to newborns — even surgery on them without benefit of pain-killing anesthetics — because of deep prejudices they had acquired in medical training.

Obstetrical Hallucinations and Delusions

In the United States, where nearly everybody is now born in a hospital, we all know that babies are typically born crying, kicking, and screaming. Why? This pertinent question is neither raised nor answered by doctors, and the crying goes on and on, decade after decade. How do obstetricians react to crying babies? Typically, they smile and congratulate mothers on what a strong voice their baby has. What are they thinking? My guess is they are thinking that everything is fine! If they would explain this aberration, they would probably say the baby is showing healthy "reflexes," but *not* having a personal *experience*. Baby expressions (whatever they might be) lacked mind and meaning, and therefore were not valid communications. Consequently, there was no point in listening to a baby or trying to interpret its sounds and gestures. There was certainly no reason to make changes in obstetrical protocols!

It took me awhile to figure out that what we are looking at here was delusional thinking based on false science. Obstetricians during birth events appear lost in hallucinations about a baby they hold tightly in their imagination, but it is not the baby in the room. The hallucinated baby is the one they have come to believe in and prefer to deal with, but this means they are out of touch with the actual baby for whom they are caring.

Where is reality here? The babies handled by doctors and nurses are (from my point of view) definitely unhappy, frustrated, and angry, as they are twisted, punctured, pulled, forced, and pressured; they are *cold, in pain, and afraid*! They are, in fact, protesting with all their might through dramatic body language everyone can plainly see, and through piercing sounds everyone can hear. Through all this clamor and commotion, professionals manage to hold to the belief that what they are seeing is not what it seems to be and what they are offering newborns is "the best care" ever invented for babies.

Historically, I think it is also fair to say that obstetricians had parallel delusions about the mothers in their care. Mothers, they asserted, are ill-equipped to give birth safely and needed constant "management." Female bodies are not only unreliable but dangerous. After delivery, women could not be trusted with their own babies, so babies were taken from them and cared for by professionals. For half a century, even mother's milk was under suspicion and hospitals recommended a manufactured product instead.

Where is reality here? Factually, mothers have been equipped for birthing since the beginning of the human era. Similarly, mother's milk — real miracle that it is — has sustained virtually all human babies in the critical months and years following birth. And, speaking in practical terms, this perfect milk is free, sanitary, alive, comes warm, and is available on demand day and night. Beyond that, however, mothers and babies are physically and emotionally interdependent and derive mutual benefits from being together before, during, and after birth in what can be called an inspired synchrony of timing and hormonal interaction. How did the professions dealing with mothers and babies drift so far into a World of Illusion?

For most of the twentieth century, two of the worst delusions shared by medicine and psychology were that babies had no perception of pain and no memory to record their experiences. These delusions spread a pall of unnecessary suffering over babies for most of the twentieth century. They deserve special mention here.

Real Memory

Renowned psychiatrist, Sigmund Freud, who had worldwide influence both during and after his lifetime, invented a memory theory called "infantile amnesia," which held that none of us had the brain development necessary for personal memory until around two years of age. This theory, strongly embraced in psychiatry, obstetrics, pediatrics, and surgery caused incalculable harm to babies from 1916 when it was first announced until 1996 when it was thoroughly discredited by experimental research. This long period of time can now be seen as "80 years of scientific amnesia." In the end, leading scholars declared that Freud's *grande illusion* was invented to explain something that did not really exist. The tragedy was that this theory obscured the true nature of infants and dissuaded both parents and professionals from giving appropriate care to babies!

In the new millennium, human memory can no longer be described by the physical boundaries made familiar by classes in anatomy and neuroscience. Babies have proved sensitive, aware, and vulnerable at all ages. Time has revealed they accumulate memories and learn from experience in ways we wish they wouldn't. They remember things that are difficult to explain, yet they learn in all the ways that learning is tested. Above all, they know if they are wanted or not, sense who to trust, and against all the odds, they can report out-of-body and near-death experiences in the womb, which had profound consequences in how they lived their lives.

Babies start learning languages in the womb, but a growing literature shows they are already communicating telepathically with mothers and fathers long before language. In fact, the latest information reported by hundreds of parents reveals intense telepathic communication before conception, that is, before they have any physical body at all.[2] All these empirical discoveries completely overflow the boundaries of the "brain matter" paradigm that dominated our thinking through most of the twentieth century.

Many of you know of my personal contribution to this paradigm shift in proving experimentally that birth memories were reliable and *not* fantasies.[3] The fuller story of birth memory can be read in *The Mind of Your Newborn Baby*[4] (third edition), which is presently circulating in 12 languages. Nevertheless, many professionals working with babies still live in the broken paradigm of "infantile amnesia." In a forthcoming book, *Windows on the Womb: Your First Nine Months*,[5] I will complete the case for a larger view of human memory — that it is *innate* rather than "developmental." Acceptance of this new psychology of consciousness in babies has the potential to redefine pregnancy and parenthood and to change how professionals work with babies from conception onward.

Real Pain

Medical doctors went through a long period of denying the significance of baby pain, even after discovery of ether anesthesia in 1846.[6] As medical historian Martin Pernick has pointed out, babies were not counted among those who *needed* anesthesia, so baby surgeries were done without painkillers. As late as 1989, a professor of pediatrics in a major US medical college wrote in a journal article on neonatal pain that "we know virtually nothing about whether there is memory of infant pain."[7] At that time, probably most pediatricians and obstetricians comforted themselves with the delusion that pain was not a factor in their work with prenates and newborns. Indeed, in many locations in the world, pediatric surgeons had operated on babies without anesthesia over a period of 140 years (from 1846 to 1986) because they believed anesthetics were unnecessary and dangerous. Denial of infant pain was one of the biggest mistakes in the history of medicine.

This dark, long, and mostly secret era of medicine began to end in 1986 because of two developments, one public and the other scientific. Parents who had lost their neonates to surgery without anesthesia went public with their story and attracted media attention for a protracted period of public debate.[8] Eventually, public opinion won out and medical guilds promised to give babies the same consideration in regard to anesthetics for surgery as they give to other patients.

The scientific revolution was initiated by pivotal research on infant pain in PhD research at Oxford University in England by Indian physician K.J.S. Anand.[9] Anand's research proved that baby surgery *with* anesthesia was definitely beneficial compared to the same surgery done *without* anesthesia, which he showed was harmful and sometimes fatal. In the 20 years since his Oxford research, Dr. Anand has continued to lead the world toward a fuller scientific understanding of the many dimensions of infant pain and the necessity of changing the way infant pain is treated. To review his extraordinary contributions, one need only do a search in Pub Med for "K.J.S. Anand" and 50 research reports will tell you how far we have come in smashing the tightly held delusions of medical practitioners that infants had no pain.

Nevertheless, a recent probe in 2006[10] reveals how medical protocols continue to overlook the significance of infant pain. In this study utilizing sophisticated brain-scanning, researchers focused on premature babies before, during, and after they underwent *routine* blood tests using a heel lance. Maria Fitzgerald and colleagues in England used near-infrared spectroscopy to measure the blood supply and oxygenation in the brain of babies between 25 and 45 weeks conceptual age. They demonstrated that surges of blood and oxygen were reaching the somatosensory cortex, an area that has been linked to pain sensation in adults, and concluded that these very young babies were experiencing true pain, not just reflex reactions. In closing their report, the team noted that in neonatal intensive care units today, babies typically receive an average of 14 procedures per day, "many of which are considered by clinical staff to be painful."

Contemplate for a moment, the potential consequences of all the unacknowledged — and routinely inflicted — pain in neonatal intensive care units in the 40 years since 1967, when the first of thousands of units in the US and the world opened for business at Yale University. This intrinsically painful mode of treatment continues today as the standard approach to saving and healing our most immature and fragile babies.

The Circumcision Crusade

In the light of all the above facts about how professionals engaged in a delusional *shrinking* of babies in the twentieth century — effectively robbing them of their smiles, cries, emotions, pains, memories, and what they were learning — let us turn attention to the practice of male and female circumcision. Here we confront compounded delusions originating in ancient tribal practices in the Middle East and Africa, dating back for *thousands* of years. Tragically, these rituals, which ran free and unchecked in pre-scientific cultures driven by fear, prejudice, and orthodoxy, continue in vogue in certain countries today, driven by the same fear, prejudice, and orthodoxy. Circumcision today perpetuates and institutionalizes a righteousness that permitted authorities of earlier times to dictate tribal markings and sexual mutilations and to force parents to comply. It is only recently in human history that genital cutting has been identified as "sexual abuse."

In the United States, it is especially ironic that, in the nineteenth and twentieth centuries, the rapid rise of circumcision was led by physicians who claimed the mantle of science with backing from formal medical education, textbooks, official journals, and membership in a guild. In those days, women were denied access to medical education and male physicians despised and opposed the ancient role of midwives at birth. Medicine became the new orthodoxy in the field of health, gained status in the public eye, and won the favor of legislative bodies passing out rights and privileges. Doctors were ranting about masturbation in the 1860s and declared that foreskin was detrimental to health.

Circumcising physicians, themselves moralistic and dogmatic, found a willing ally in the maverick millionaire John Harvey Kellogg, whose tract, *Plain Facts for Old and Young* (1888),[11] was a household item praising the virtues of circumcision. Mr. Kellogg seized upon circumcision as a way of stopping what he called "the evil of masturbation." (Now *there* is a serious problem!) With no real evidence that it was actually evil or had evil effects, Mr. Kellogg continued on his passionate crusade with the hearty support of physicians who had even more elaborate "scientific" delusions about circumcision as a means of preventing or curing a growing list of diseases!

Delusions are surely an amazing phenomenon. As a psychologist, when I first met the term "delusion," it was a highly specialized disorder, discreet from other derangements, and rare. To find one, you had to go to a lunatic asylum. Now, it seems, delusional thinking is everywhere you look. It struts and poses without apology, has high status, wears uniforms, claims divine inspiration, and spreads like the flu.

In the 1890s, doctors asserted that the intact penis was associated with a garden variety of "nervous conditions," tuberculosis, and dozens of other human afflictions. Thus began a surgical stampede to circumcise urban males in America, with a new aura of science covering the same old crime — willful injury and robbery of functional sexual body parts of babies and children by tribal elders.

The perverse extension and resilience of this long-lasting movement is remarkably illustrated in Frederick Hodges' *A Short History of Circumcision in North America: In the Physicians' Own Words* (1996).[12] Hodges' chart begins in 1860 when only 0.001% of the urban male population was circumcised. Twenty-eight years later, the percentage had risen to 15% and 40 years later (1900) to 25%. With unceasing claims of benefit, the percentage reached 50% in 1920 and came to a peak 50 years later, in 1971, when 90% of urban males were being circumcised.

From 1900 to 1935, doctors claimed in medical publications that circumcision would increase sexual power and control, prevent syphilis, dyspepsia and diarrhea, convulsions, epilepsy, prolapse of the rectum, dropsy, crying in infants, and, yes, hydrocephalus. Circumcision could prevent adolescent rapes and promiscuity, marital separations and divorce caused by "unnatural passion," and penile cancer due to chronic irritation of foreskin. In 1949, doctors in medical journals made sweeping assumptions that Negroes were more promiscuous and had more venereal diseases because fewer of them were circumcised. Similar authoritative statements were published that cancer of the prostate, cervix, and penis occurred in groups with low circumcision rates. Claims in the 1980s proclaimed that circumcision decreases urinary tract infections, and could protect from AIDS.

These extravagant claims — none of which finally survived empirical scrutiny — fueled the delusional mania for circumcision for over 100 years, until 1971, when credibility finally began to crumble. A medical textbook used at that time reminded medical students that most sensible parents would welcome circumcision "as a way to avert masturbation." But it was too late. Twenty-five years later — with a lot of help from the National Organization of Circumcision Information Resource Centers (NOCIRC) — the percentage of urban males being circumcised had fallen from 90% to 60%.

Injury and Harm to Females and Males

Genital cutting of males and females has been international in scope for millennia. In the twentieth century, remnants of female cutting have been recorded in the United States as late as 1977, when Blue Cross/Blue Shield still paid for clitoridectomies. The reality of this era is personalized in the autobiographical book, *The Rape of Innocence* (2006),[13] by Patricia Robinette. In the United States, a law against female genital mutilation went into effect on March 30, 1997.

According to UNICEF at least 100 million women, largely in Africa, have been genitally disfigured in childhood. A WHO study group led by Emily Banks from Australia National University (2006)[14] has shown that women with female genital mutilation (FGM) are more likely than other women to suffer a stillbirth or early neonatal death. The group studied the obstetrical outcome of 28,393 women who attended one of 28 obstetrical centers in Burkina Faso, Ghana, Kenya, Nigeria, Senegal, and Sudan. They used a WHO rating system to evaluate severity of mutilation. FMG-I is removal of the prepuce or clitoris or both; FMG-II is removal of clitoris and labia minora; and FMG-III is removal of part or all of the external genitalia with stitching or narrowing of the vaginal opening. Risks tend to rise with more extensive mutilation.

The team found that women with FGM were more likely to lose their baby during the perinatal period than women without FGM. Other adverse obstetrical outcomes included cesarean section, hemorrhage, episiotomy, and an extended hospital stay. They concluded that adverse obstetrical and perinatal outcomes can be added to the known harmful immediate and long-term effects of FGM. Considering that the purpose of FGM is to maximize the value of the girls as brides, the actual outcome of FGM (stillbirth and early neonatal death) was actually the kiss of death.

A study in Egypt (2007)[15] looked at the impact of genital cutting on the health of newly married women. In a random group of 264 newly married, they compared circumcised and non-circumcised women regarding long-time health problems. In Benhe City, Egypt, the circumcision group was 76% of the sample. All lived in an urban area. Circumcised women had significantly more dysmenorrhea, marital and sexual problems, obstetrical complications, anxieties, and phobias. The authors concluded, "Grave problems of circumcision may last throughout life ..." and were particularly disruptive at the time of consummation of marriage and the time of childbirth. The consequences were opposite from what the families were expecting.

The outcome for males in those cultures where circumcision is commonly imposed on newborns or children are also contrary to the many advantages promised by doctors. The paper by Boyle, Goldman, Svoboda, and Fernandez (2003),[16] "Male Circumcision: Pain, Trauma, and Psychosexual Sequelae," provides a realistic and thorough analysis of the consequences of male circumcision, beginning with the immediate tissue loss caused by cutting around the penis to amputate part or all of the foreskin. And this is only the first irreversible damage to body and psyche. (For the full story of the harm wrought by circumcision, please study this article.)

Although perhaps 85% of the world's adult males remain genitally intact, an estimated 650 million males alive today have been circumcised. In the US, over a million baby boys are added to this group every year. The excruciating pain of this surgery has been measured in every conceivable way, leaving no doubt about the intense shock and suffering the baby endures. The purpose of the event may vary in different cultural groups, but in the United States, the physicians who promoted and popularized the practice intended to make a deep impression on the newborn, weaken the penis, and repress sexual feelings and interests. In contemporary language, this was a flagrant violation of human rights, abusing children in no position to defend themselves or to give consent.

Up until the year 2000, most babies were circumcised without benefit of pain-killing anesthetic while they were restrained and helpless. Although many doctors in those days still believed babies could not truly feel pain, interpret its meaning, or remember and learn from the experience, we know today beyond doubt the experience was and is a traumatic ordeal physically, emotionally, and mentally. Research since that time has shown that circumcision trauma permanently lowers the pain threshold for subsequent experiences of pain. Clinical work with babies, children, and adults often reveals long-lasting post-traumatic stress disorders, anxieties, and phobias created by brutal early experiences of surgery and circumcision. In addition to personal suffering, therapists have realized devastating breeches of trust and bonding failures between babies and their mothers and fathers. Trauma distorts behavior, personality, self-esteem, and the wounded penis will be carried into all future sexual relationships.

We can thank NOCIRC for designing research that directly challenges the pretensions and delusions of circumcision advocates who promised better sex and freedom from a cornucopia of diseases by removing foreskin. Long in the planning stage, this elegantly simple research, completed by a team of seven doctors and nurses, may go a long way toward taking the hoax out of circumcision. The title is: "Fine-touch pressure thresholds in the adult penis" (2007).[17] The objective was to map the fine-touch thresholds of the adult penis in circumcised and intact men and compare the two populations.

One hundred sixty-three subjects were enrolled. The fine-touch sensitivity of 19 locations on the penis was measured using Semmes-Weinstein monofilament touch-test sensory evaluators to create a map of penile sensitivity. Results show that the most sensitive regions of the intact penis are exactly those removed by circumcision. When compared with the most sensitive area of the circumcised penis, several locations on the intact penis (missing from the circumcised penis) were significantly more sensitive. The glans in the circumcised male is less sensitive to fine-touch pressure than the glans of the intact male. The most sensitive location on the circumcised penis is the circumcision *scar* on the ventral surface. Five locations on the intact penis were more sensitive than the most sensitive location on the circumcised penis.

Despite the controversy over the long-term impact of male circumcision, no thorough, objective, quantitative studies measuring the long-term sensory consequences of infant circumcision have hitherto been reported. The present study provides the first extensive mapping of the fine-touch pressure thresholds of the adult penis. The many partial attempts to extend our knowledge in this area are expertly analyzed in this paper. *In conclusion, circumcision removes the most sensitive parts of the penis and decreases the fine-touch pressure sensitivity of glans penis. The most sensitive regions in the uncircumcised penis are those parts ablated by circumcision.*

These clear facts challenge both doctors and parents. How are we to understand a medical system that claims to offer miraculous benefits while doing permanent damage to newborn boys? On the other hand, how are we to understand the delusions that drive otherwise normal parents to hire doctors to perform miracles that are a hoax?

Civilization: Dreams and Delusions

A news reporter once asked Albert Einstein what he thought of Western Civilization? His quick reply was, "I think it would be a good idea." I, too, think civilization would be a good idea, but I am beginning to see that our fantasies, delusions, and other false beliefs may prevent us from getting there. In approaching the larger subject of civilization, I think there are things we can learn from the debacle of the circumcision crusade in America — particularly because babies are so deeply involved in each venture. Note, for example, the truths that emerge from the following declarations relating to familiar old delusional ideas about circumcision.

1. Injuring a normal penis does not improve its function but, in fact, degrades it. Common wisdom about "improving" a penis would be to just leave it alone!
2. Inflicting any form of severe pain on newborns does not make them more joyful, confident people. To the contrary, shock, brutality, and torture of babies actually makes them anxious, fearful, and vengeful. The big question to ask is, What strategies would help to create joyful, confident babies, children, and adults?
3. Complicity of parents in causing the wounds of circumcision cannot be hidden from victims because the unconscious memory still holds them, and will erode the foundations of trust. Important priority: We must learn how to establish trust.
4. Robbing babies of healthy body parts is a violation of human rights and will not improve character or increase family solidarity. The challenging question is: What *would* improve or increase family bonding?
5. Being victimized by a surgery performed without your consent is not an experience of democracy or an exercise in positive psychology. If this is what we want, we could have it by just omitting the surgery.

Consider what can be learned from de facto experiments of much larger scope that affect not only individual persons but whole societies. In the twentieth century, huge public health "experiments" were conducted without any measurements, controls, evaluations, or debates. Not scientific in any sense, the suffering involved was pointless and inhumane. Circumcision was only one such "experiment." Others included these: feeding babies manufactured "formula" instead of mothers' milk and replacing breasts with bottles; the cascading of vaccinations against numerous diseases; pollution of drinking water, land, and air with pesticides; the mass production of automobiles and the reorientation of communities around vehicles burning fossil fuels; the general use of toxic mercury fillings in dental offices; cluttering the oceans of the world with durable plastic trash; radiant fallout from nuclear power plants and waste piles; and, of course, one of the biggest potential problems, global warming. The list goes on, affecting not only neighborhoods but continents, and capable of degrading the whole planet. Such problems can nullify all efforts to achieve "civilization." ("A nuclear bomb can ruin your whole day!")

Such problems are sobering and force us to rethink the very nature of civilization and the strategies appropriate to getting there. What is civilization, after all? More paved streets, more cars, higher buildings, or a thousand Walmarts in China? We might have difficulty arriving at an agreement about civilization as a place, or what would make it civilized. Once upon a time, the junction of the Tigris and Euphrates Rivers was considered "the cradle of civilization" but that place has changed. It might be more helpful to focus on civilization as "civility," a quality of life, relationships, freedom, sharing, and cooperation? Or, civility as a principle of using all resources for the common good? All this sounds good to me as a psychologist interested in wellness, human growth, altruism, and realizing that the human race is all one family. But I am thinking, if this is where we are headed, we will very much need to apply what we have learned in prenatal psychology about real babies. Babies are bound to play a critical role in both family life and planetary life. ("Womb ecology becomes world ecology.")

Who Are the Gatekeepers?

If ultimately, the quality of a civilization depends upon the quality of the people participating in it, we will at least need a critical mass of people with the skills, attitudes, and virtues needed to hold together the increasingly large Global Family of Man. Personally, I am not sure about institutional structures as the gatekeepers of civilization, despite their inclinations to assume they are. My own view is that the institutions themselves — whether they are governments, schools, media networks, scientific and professional guilds, churches, libraries, or charities — inevitably depend on the wisdom and maturity of the people who, at any given time, make up these institutions.

One of the glaring signs of institutional failure during the twentieth century is the delusion that violence solves all problems. The record is appalling, decade by decade, including the first and second World Wars, Nazi expansionism and the Holocaust, Stalinism, civil terror in Communist China and a genocide in Tibet, brutal dictatorships in South America, the cruelty of apartheid in South Africa, wars in Korea, Vietnam, the Middle East (where circumcising cultures chronically clash), Yugoslavia, Rwanda, Afghanistan, Iraq, and the new brand of terrorists, women and children suicide bombers.[18] In all this frantic activity, there were never enough people in the right places and at the right times to find solutions to the problems and make the wars unnecessary. Wars are still waiting to be understood. Meanwhile, as the actor Peter Ustinov puts it, "Terrorism is the war of the poor and war is the terrorism of the rich." In the US, we name bombs "The Peacemaker" and are proud of "shock and awe."

Delusions aside, someone has suggested that war is not problem-solving but a manic-depressive illness of society. The manic phase is seen in recruiting, training, shipping soldiers around the world, then shooting babies, raping mothers, enlisting children to shoot others — all of which is very exciting and savage! The depressive phase is paying billions to rebuild the enemy's infrastructure: restoring water, electricity, transportation, and communications, then facing and paying off the massive loans that financed the war effort. Who can step in and save us from our collective folly?

The Real Power of Loving Families

My best guess about where to look for the people who might actually know how to build a real civilization is in loving families. This is surely oversimplified and too idealistic to be taken seriously, but frankly, these families may be our only hope. Considering the collective weaknesses and shortcomings of the world's parents, this may be the scariest theory of civilization ever brought forward. Parents could very well let us down. Some families are notoriously dysfunctional and miserable, others are a breeding ground for violence, sickness, and depression. Granted, the odds are not good that these families will produce the inspired, healthy, and creative people that are needed in a civilized world — although some real life stories do prove they *can* sometimes beat all the odds. When a child succeeds under tough circumstances like that, I tend to think that the explanation doesn't lie in the shortcomings of the parents but in the extraordinary character of the child.

But, let's take a closer look at the "odds." What would be the odds of producing healthy, happy, and friendly children in a society where 100% of the girls suffered genital mutilation and 100% of the boys were robbed of their foreskins? Would you expect an improvement in this imaginary society if only 50% of the girls and 50% of the boys were mutilated? Finally, what would you think of the odds if *none* of the girls and *none* of the boys were circumcised at all? This illustrates what odds really mean, and this brings me back to the power that resides in the loving family: *loving makes a huge difference!*

If babies have taught us anything, it is that they are little sponges, soaking up the environment of the mother and father; they are constantly involved, constantly learning, and very impressionable. While they are having these learning experiences, they are literally forming their brains, their emotions, and their ideas about themselves and establishing what might be called their default "factory settings" for life. Another way to speak about this process is that the *parents* are in a real sense the architects of the brain that is growing in the child because the parents are controlling most of the input, including the fuel needed for growth. The truth is this: in a loving family the input is basically loving, while in a hostile and crazy family the input is mostly hostile and crazy.

We are talking about the quality of the environment here, including all the important elements like nutrition, affection, sound, language, other higher levels of communication, including intention and purpose, modeling, and all other types of learning and "vibes."

Out of this interactive milieu will emerge sexual orientation, behavioral tendencies, and personality traits, one of which will have especially profound implications for civilized behavior; it is called your "dispositional signature." Disposition accounts for stable and recognizable aspects of individuality, the style of a person's adjustment to and engagement with the social world, including how a person does things, typically thinks, and usually feels. It tends to be a robust predictor of important life outcomes like work performance and occupational success, quality of social relationships, psychological well-being, and even longevity.[19] What the world needs is lots more people with a disposition to be friendly, cooperative, and helpful — people who are disposed to be peaceful and civil rather than angry, prejudiced, critical, and troublesome. A good disposition is shaped in a loving family.

On the other hand, there is risk of developing a violent disposition. Violence is home-grown, and turns into one of the worst problems we face in today's fragile world. Violence is learned from experience, especially from watching, hearing, or experiencing it *up close* from parents. Later, television violence spreads the confusion far and wide and makes it seem normal, but long before television, babies have sensitive radar for violence, whether it happens at conception, during pregnancy, at birth, or any other time. They memorize it, sense it, try it out, and react to it by becoming afraid or aggressive. Violence grows with practice; the more you act violent, the more violent you become until you are better at it than anyone else. Domestic violence becomes public violence, and public violence can multiply itself, a phenomenon often seen on the evening news.

Something else super important to civilization and is home-grown is your "*primal health system*" — the system that protects your health for the rest of your life. Experts are now telling us that this system is constructed in the period between conception and your first birthday. The diseases you are likely to develop in life and may die from can often be predicted, knowing the conditions you faced during this critical time period. Therefore, your quality of life — and the health of your society — depends on the quality of your primal health system.[20]

The bottom line is that civilization requires a critical mass of peaceful, intelligent, healthy, sociable, good-humored people to make it happen. The babies that come into families are the future population of the world, and those surrounded with love will likely be the harbingers of a great society.[21] I can't think of any other place from which civilized people could come.[22]

Warnings and Conclusions

- Violence and pain are not a secret back door entrance to civilization.
- Cutting off foreskins and mutilating erotic organs is violence, not a health policy.
- Delusions of parents and birth professionals will not advance human rights or human life.
- Civilization is not a place, institution, army, or shopping mall; it's a way of being human.
- Basic skills, values, and principles of civilization are learned in loving families.
- It may be the scariest theory yet, but without loving parents, we may never achieve a true civilization.

References

1. Dick-Read, G. (4th ed.) Childbirth without fear. New York: Harper Colophon Books 1944/1978.
2. Hinze, S. Coming from the light. New York: Pocket Books. 1994. Hallett, E. Soul trek: Meeting our children on the way to birth. Hamilton, MT: Light Hearts Publishing. 1995. Hallett, E. Stories of the unborn soul. Hamilton, MT: Light Hearts Publishing. 2002. See also the column: Communication before conception which can be viewed at www.Birthpsychology.com/lifebefore/concept.html.
3. Chamberlain, DB. Reliability of birth memory: Observations from mother and child pairs in hypnosis. Journal of Prenatal and Perinatal Psychology and Health 1999;14(1–2):19–29.
4. Chamberlain, DB. The mind of your newborn baby. Berkeley, CA: North Atlantic Books. 1998.
5. Chamberlain, DB. (Forthcoming) Windows on the womb: Your first nine months. Berkeley, CA: North Atlantic Books.
6. Pernick, MS. A calculus of suffering: Pain, professionalism and anesthesia in nineteenth century America. New York: Columbia University Press. 1985.
7. Marshall, RE. Neonatal pain associated with care giving procedures. Pediatric Clinics of North America 1989;36(4):885–903.

8. Harrison, H. Letters. Birth 1986;13(2):124. Lawson, J. Ibid.126. Also: Lawson, J. The politics of newborn pain. Mothering Magazine 1990;(Fall):41–46.
9. Anand, KJS. Metabolic and endocrine effects of surgery and anesthesia in the human newborn infant. PhD Dissertation, Oxford University Press (England). 1985.
10. Slater, R, Cantarella, A, Gallella, S, Worley, A, Boyd, S, Meek, J, and Fitzgerald, M. Cortical pain responses in human infants. Journal of Neuroscience 2006;26(14):3662–3666.
11. Kellogg, JH. Plain facts for old and young. Burlington, IO: F. Segnur & Co. 1888.
12. Hodges, FM. A short history of circumcision in North America in the physician's own words. http://www.noharmm.org/docswords.htm.
13. Robinette, P. The rape of innocence: One woman's story of female genital mutilation in the U.S.A. Eugene, OR: Aesculapius Press. 2006.
14. WHO Study Group on female genital mutilation and obstetric outcome; WHO collaborative prospective study in six African countries. Lancet 2006;367(9525):1835–1841.
15. Einashar, A and Abdelhady, R. The impact of genital cutting on health of newly married women. International Journal of Gynecology and Obstetrics 2007;97(3):238–244.
16. Boyle, G, Goldman, R, Svoboda, J, and Fernandez, E. Male circumcision: Pain, trauma and psychosexual sequelae. Journal of Health Psychology 2002;7(3):329–343.
17. Sorrells, M, Snyder, J, Reiss, M, Eden, C, Milos, M, Wilcox, N, and Van Howe, R. Fine-touch pressure thresholds in the adult penis. British Journal of Urology, International 2007;99: 864–869.
18. Brekhman, GI and Fedor-Freybergh, PG (Eds.) Phenomenon of violence (from domestic to global): View from a position of prenatal and perinatal psychology and medicine. Haifa: NEWS AGALIL 2005. p. 185.
19. McAdams, DP and Pelt, JL. A new big five: Fundamental principles for an integrative science of personality. American Psychologist 2006;61(3):204–217.
20. Odent, M. Primal health: Understanding the critical period between conception and the first birthday. East Sussex: Clairview Books. 2002.
21. Grille, R. Parenting for a peaceful world. Alexandria, New South Wales, Australia: Longueville Media. 2005.
22. Witt, K. The attuned family: How to be a great parent to your kids and a great lover to your spouse. Santa Barbara, CA: Santa Barbara Graduate Institute Publishing. 2007.

Reconsidering 'Best Interests'

Male Circumcision and the Rights of the Child[1]

Marie Fox and Michael Thomson

Abstract Recently the ethics of infant male circumcision has generated a considerable debate in bioethics journals. In previous publications we have sought to argue that, by contrast, healthcare lawyers have unjustifiably neglected the topic, accepting a dominant characterization of male circumcision as a standard and benign medical practice, which parents can choose for their children free of legal scrutiny. In this paper, we seek to problematize both the way in which male circumcision is debated in the ethics literature and how it is constructed as a non issue for healthcare lawyers. We are concerned here particularly with the cost/benefit analysis that has underpinned professional guidance and court decisions on the legitimacy of male circumcision. We argue that how these costs and benefits have played out in the ethico-legal debates and assessments as to what is in the best interests of the child are highly problematic.

The debate on male circumcision, in contrast to that on female circumcision is marked by an absence of statute law. There are also few reported cases. In the absence of clear legal authority, 'soft' law sources, particularly guidance issued by professional medical bodies, assumes enhanced significance. In the United Kingdom, the British Medical Association issued revised guidance as recently as 2003, but it is currently being re-considered. The current General Medical Council Guidance dates from 1997, while in 2001 a Joint Statement on male circumcision was issued by the British Association of Paediatric Surgeons, The Royal College of Nursing, The Royal College of Paediatrics and Child Health, The Royal College of Surgeons of England, and the Royal College of Anaesthetists. A common theme of this guidance is the way in which it constructs the issue of male circumcision as a legitimate parental choice—a position which we argue is in need of review.

In this paper we adopt a comparative approach in order to examine guidance on male circumcision issued by professional medical associations in countries which share a similar tradition of circumcision to the United Kingdom. Thus, we examine guidance issued in the United States, Canada, Australia, and New Zealand. Our focus is on professional guidance, that adopts a more progressive and less tolerant approach to the issue of elective neonatal circumcision. Thus, we seek to explore

School of Law, University of Keele, UK

G. Denniston et al. (eds.), *Circumcision and Human Rights*,

what the medical profession in the United Kingdom and the United States could learn from guidance issued by the Fetus and Newborn Committee of the Canadian Paediatric Society in 1996 and in a Policy Statement on Circumcision issued by the Paediatric and Child Health Division of the Royal Australian College of Physicians in 2002.

Introduction

In this essay, we aim to consider how effectively United Kingdom (UK) law and policy protects the rights of the child, given its willingness to allow parents to choose non-therapeutic neonatal male circumcision. We adopt a children's rights framework for this analysis in recognition of the prevalent discourse, which has characterized UK health law pertaining to children since the late 1980s.[2] As Andrew Bainham notes:

> [T]he one feature which best characterises late twentieth century development in the law relating to children...might well be the shift away from law's almost exclusive concentration on the protection of the individual child to the recognition of the interests, indeed rights, of children as a class or group.[3]

However, perhaps inevitably, the development and protection of children's rights in law has been somewhat uneven. We would suggest that this is particularly evident when one explores legal approaches to the genital cutting of children. In contrast to the legal prohibitions concerning female genital cutting,[4] UK law and policy has adopted a permissive approach to male genital cutting. Analyzing the limited case law on the issue in the UK, along with the professional guidance issued by the British Medical Association (BMA), we argue that the prevailing ethico-legal view that circumcision in these circumstances is a matter of parental choice is incompatible with current standards of child protection evidenced in other areas of health-care practice. In contrast to some commentators, however, we argue that it is strategically important to stop short of labelling this practice a form of child abuse.[5]

Given the United States (US) focus of this collection, we start by considering the current incidence of neonatal circumcision in the UK, before addressing the current case law and the most recent professional guidance issued by the BMA. This law and guidance are, to a large extent, delineated by the "best interests" test. The main objective of this essay is thus to question the extent to which this test adequately protects the rights of the child. Instead, we advocate a "needs" and "harm"-based approach to this issue, suggesting that such discourse holds the potential to more effectively promote the bodily integrity of those who are not yet capable of consenting to medical interventions.

Whilst we conclude that construing non-therapeutic neonatal circumcision as a matter of parental choice is incompatible with current standards of protection generally offered to children in the healthcare setting, we argue that tackling this practice is best done through dialogue and education, rather than legislative prohibitions.

Such an approach, in our view, is preferable to the punitive way in which female genital cutting has been approached in the UK and other jurisdictions.

Non-Therapeutic Circumcision in the UK

The US and the UK share a somewhat disjointed circumcision history. Most other European and industrialized nations — with the exception of Canada and Australia — do not exhibit a similar history of high rates of routine neonatal circumcision.[6] However, this common history is disjointed, in that the prevalence of circumcision in both countries has not endured. As in the US, in the early decades of the last century, the majority of newborn males in the UK were circumcised. The overall incidence — medical and social — of circumcision was very clearly stratified along class lines. Thus, based on Army records, it has been estimated that before World War II 50% of working class and 85% of upper class men in England were circumcised.[7] In the early 1930s, 35% of these procedures were for medical reasons.[8] Whilst US rates remain (unevenly) high, in the UK the numbers, though significant, are now comparatively small. The NHS started to provide operations in 1948. Following the publication of Gairdner's influential report in 1949 contesting the necessity of neonatal circumcision,[9] the incidence of non-therapeutic (routine) circumcision declined sharply.[10] It was reported to have fallen to 6% by 1975,[11] while Johnson, Wadsworth, and Wellings' survey of 7,990 British men in 1990 found that 21.9% of all men in the survey were circumcised.[12]

Whilst recorded rates vary and the accuracy of the figures is difficult to verify, more recent estimates confirm an incidence rate in single figures. In 1993, it was suggested that approximately 30,000 procedures were performed annually in the UK, most on young children.[13] In 1995, the incidence was estimated to be "less than 10%."[14] In 1997, Warren calculated that "since 1950 the cumulative circumcision rate in boys has dropped from about 30% to about 6% or 7%."[15] In 2000, it was estimated that approximately 12,200 circumcisions were performed in England in 1998 for medically indicated reasons.[16] Whilst the study notes that this accounts for a 10,000 drop in procedure levels since 1992–1993, the authors argue that two-thirds of procedures were unnecessary and resulted from the misdiagnosis of phimosis.[17] With a circumcision rate of 6.5% this assertion was supported by the 2% circumcision rate in Scandinavia.[18] More generally, the *British National Survey of Sexual Attitudes and Lifestyles 2000* found that 15.8% of males between the ages of 16 and 44 reported being circumcised. Given changing attitudes to circumcision, it is unsurprising that circumcision was highest in men born between 1956 and 1960 — that is, those aged between 40 and 44 at the time of the study. This cohort reported a rate of 19.6%. The lowest incidence was recorded in the youngest age group, 16–19, where the rate was 11.7%. These men would have been born between 1981 and 1984.[19] A recent study by Cathcart et al., charts the declining incidence of male circumcision over the period 1997–2003 and suggests that approximately 3% of British boys aged 15 are circumcised.[20] Returning to the Scandinavia rate noted above, this suggests that one in three procedures are still performed unnecessarily.

Law and Policy

Essentially UK law provides that so long as both parents are in agreement, the decision to circumcise a neonate is a legitimate parental choice with which courts should not interfere. As we have explored elsewhere, professional guidance, in common with law, exhibits an (almost) unparalleled willingness in the case of neonatal male circumcision to tolerate the non-therapeutic, non-consensual excision of healthy tissue.[21] The one exception to this is surgical interventions on children born with intersex conditions, although here there is evidence that practice is changing and the trend in favor of surgery declining.[22] Thus, infant male circumcision is characterized by an acceptance of levels of risk unimaginable in other healthcare contexts.

We have sought in earlier papers to contest a discursive pattern whereby risks of circumcising male children are downplayed through a strategy of contrasting the practice of male circumcision with that of female circumcision.[23] By way of a brief example, this technique is evident in the BMA's revised guidance of 2006, discussed below, where male circumcision is distinguished from "very harmful cultural procedures, such as female genital mutilation or ritual scarification."[24] In these earlier papers our aim has been to question how cost–benefit analyses of the harms and benefits of the practice have typically minimized the risks in this way, leading to the conclusion that they are sufficiently *de minimas* to be left to parental discretion. The process is demonstrated vividly in the following passage from two bioethicists. Notwithstanding their pledge to take the cost/benefit calculation of the procedure seriously, they conclude:

> [W]e think that neonatal circumcision cannot unequivocally be said to yield a net medical gain or loss. In other words, it is not something that can be said to be routinely indicated, nor something that is routinely contraindicated. It is a discretionary matter. The decision whether or not to circumcise a child should thus be made by the parents, who, within certain limits, are entitled to employ their own value judgments in the furtherance of their child's best interests. These limits are not exceeded in most decisions about neonatal circumcision, given the nature of the medical evidence.[25]

Yet, only through downplaying the clinical risks of the practice can the authors embrace the cultural and social benefits that might flow from its exercise. Significantly, a later and much more detailed cost/benefit analysis rooted in a comprehensive literature survey and employing statistical analysis reached the clear conclusion that:

> The perpetuation of neonatal circumcision cannot be justified financially or medically; therefore, any justification for the practice must be based on religion, culture, or aesthetics.... Currently in the United States, cultural considerations trump financial and health concerns when deciding to have a newborn male circumcised.[26]

Yet, the pattern of reasoning which inflects the Benatars' conclusion is evident not only in bioethical analyses of male circumcision, but as a tenet underpinning both the BMA's current guidance to doctors on the practice and the leading English case to have addressed the legality of circumcision.

Originally revised in 2003, the British Medical Association guidance to doctors was revisited but left largely unchanged in 2006.[27] It too foregrounds the need for a cost/benefit analysis and, significantly, within this it highlights the contested nature of the claimed benefits. At various points, evidence for the supposed beneficial effects of circumcision is described as "equivocal," "inconclusive," "not convincingly proven," "contradictory," causing "significant disagreement," "lacking consensus" and, ultimately, "insufficient," leading to the conclusion that "evidence concerning the health benefit from non-therapeutic circumcision is insufficient for this alone to be justification."[28] Yet, while this is coupled with a recognition that there are inherent medical and psychological risks in the procedure, the dominant message remains that parental beliefs should be respected, despite not being grounded in claims to health benefits:

> The medical harms or benefits have not been unequivocally proven but there are clear risks of harm if the procedure is done inexpertly. The Association has no policy on these issues. Indeed it would be difficult to formulate a policy in the absence of unambiguously clear and consistent medical data on the implications of the intervention. As a general rule, however, the BMA believes that the parents should be entitled to make choices about how best to promote their children's interests, and it is for society to decide what limits should be imposed on parental choices.[29]

As just noted, the BMA revised its guidelines in June 2006. Whilst the new guidelines acknowledge that circumcision is increasingly controversial, it nonetheless steers the same path as the preceding guidelines. However, the guidelines are interesting for a number of reasons. There is, for instance, a significant slippage in the use of the terms "non-therapeutic," "ritual," and "religious." Of specific relevance here, however, is the particular way in which, within the cost/benefit analysis, the child's *best interests* are tied to understandings of *parental* interests. There is, for example, a presumption that these interests correspond: "Although they usually coincide, the interests of the child and those of the parents are *not always* synonymous."[30] This coincidence of interests, or more accurately the manner in which the child's interests are trumped or determined by parental interests, is more explicitly illustrated in how the guidelines respond to the situation where parents may disagree over the circumcision of a child:

> If parents disagree about having their child circumcised, the parent seeking circumcision could seek a court order authorizing the procedure which would make it lawful, although doctors are advised to consider carefully whether circumcising against the wishes of one parent would be in the child's best interests.[31]

Hence, while the BMA's recognition of both the equivocal nature of the claimed benefits and the clear risks of harm (although this is subsequently minimized) may be interpreted as a progressive position,[32] it nonetheless continues to construct male circumcision as an expression of parental privilege. Indeed, in the 2006 guidelines, the child's best interests are conflated with parental interests, which appear to factor into the cost/benefit analysis. This downplays both the pain experienced by the neonate[33] and the fact that, while complication rates from routine circumcision are low, the chances of these complications being mutilatory, infective, or hemorrhagic are high.[34] Indeed, complications are potentially catastrophic, since death, gangrene, and total or partial amputation are known adverse outcomes.[35]

The guidance stresses the necessity for the agreement by both parents — which was to prove decisive in the main legal precedent. This case concerned a 5-year-old boy — "J" — who, after his parents' separation, lived with his mother. His father — a non-practising Turkish Muslim — wanted J to be circumcised so as to identify him with his father and confirm him as a Muslim. Having considered J's probable upbringing, the Court of Appeal concluded that J should not be circumcised because he was not, and nor was he likely to be, brought up in the Muslim religion. Rather he had "a mixed heritage and an essentially secular lifestyle" and was unlikely to have such a degree of involvement with Muslims as to justify circumcising him for social reasons. In these circumstances, the Court of Appeal upheld the High Court judgment by Wall J (the judge) which held that the boy was unlikely to derive any of the social or cultural benefits from circumcision that commentators such as Benatar and Benatar point to:

> [T]he mother, as J's primary carer, would find it extremely difficult to present the question of circumcision to J in a positive light, and unlike ritual circumcision occurring in the context of a Muslim family... J's circumcision would be likely to be surrounded by tension and stress.... The strained relationship between the parents and the fact that, as a circumcised child J would be unlike most of his peers, increases the risk that J will suffer from adverse psychological effects from being circumcised.[36]

Notwithstanding the positive outcome in this case, with a ruling that it was not in the child's best interests to be circumcised, we would argue that two features of this case, which echo the BMA's guidance, are problematic. The first is the re-assertion of the view that the "family" is generally the appropriate decision-maker in cases involving the welfare of young children.[37] Thus, as Sherry Colb notes, in such cases, it is only parental conflict that "allows scrutiny of practices that would ordinarily go unexamined and permits us to ask a question that we usually refrain from asking: Is circumcision in the best interests of the child?"[38] More fundamentally, however, we would suggest that invocation of the best interests test in this context is problematic. Our discomfort arises from the clear acceptance that, in cases where the family unit is intact, the assessment of the child's "best interests" is left to the parents and the courts will not intervene.

The "Best Interests" Test

In the UK, as we have noted, the notion of the child's best interests is held to determine the parameters of medical care and treatment. The BMA guidance from 2006 recognizes that circumcision has medical and psychological risks and that it is essential that the procedure is carried out only where it is demonstrably in the child's best interests. Whilst recognizing the importance of respecting parental rights, the guidance pays considerable attention to the assessment of "best interests." It states unambiguously that parental preference alone is insufficient to justify circumcision — that preference must be explained and justified with reference to the child's interests. The responsibility to demonstrate that non-therapeutic circumcision is in a particular child's best interests therefore falls to his parents.

The guidelines also stress the place of religion and culture in any assessment. Relying on the BMA publication, *Consent, rights and choices in healthcare for children and young people*,[39] they provide a checklist of factors that may be relevant to a best interest assessment for non-therapeutic circumcision:

- The patient's own ascertainable wishes, feelings and values
- The patient's ability to understand what is proposed and weigh up the alternatives
- The patient's potential to participate in the decision, if provided with additional support or explanations
- The patient's physical and emotional needs
- The risk of harm or suffering for the patient
- The views of parents and family
- The implications for the family of performing, and not performing, the procedure
- Relevant information about the patient's religious or cultural background
- The prioritizing of options which maximize the patient's future opportunities and choices

This list offers an insight into the contours of contemporary "best interests" assessments, since judges in a number of cases have also endorsed the notion that a checklist should be used.[40] Whilst there is much to be said in this regard, we wish to focus on two aspects of this checklist in the remainder of this essay. Ignoring the first three criteria, which are clearly irrelevant to the case of neonatal circumcision, it is evident that the checklist weighs the individual interests of the infant (highlighting physical and emotional needs, the risk of harm and suffering, and future opportunities and choices) against what might be understood as collective family or cultural interests (views of parents and family, impact on the family, and religious or cultural beliefs). As such, the checklist and the wider guidelines accept that the recognized risks of circumcision can be justified by the (potential) social benefits of circumcision (or disbenefits of not circumcising).

Given the quantifiable risks of non-therapeutic circumcision (and the extent to which any adverse outcome can be catastrophic for the minor), and the unquantifiable nature of the benefits that are claimed to flow from this procedure, we wish to focus in particular on the inclusion in the checklist of the reference to the child's "physical and emotional needs." An emphasis on "needs" rather than "interests" may, we argue, provide a more focussed approach to protecting children's rights. A focus on needs is also in line with recent questions that have been raised regarding the suitability and effectiveness of the best interests test.

Problematizing the Best Interests Test

As we have noted, the best interests test remains firmly entrenched as the appropriate standard in cases involving children.[41] Although there are indications of late that its power is waning[42] and its application to young children has been criticized by some leading medical law commentators,[43] nevertheless, appellate-level judgments in the UK have continued to endorse it.[44] A major problem with the prevalence of this test is the inability of the "best interests" formulation to offer a meaningful guide to courts. In most instances, such as cases of parental refusal of pediatric treatment, it effectively operates as a cloak for the courts to uphold medical opinion. As Shaun Pattinson notes:

> The courts will act as final arbiters and will apply the best interest test to settle any dispute. Medical opinion will evidentially be a key factor in such cases. The courts have not yet deviated from medical opinion as to whether treatment should be withdrawn from a sick child, at least where that opinion has been unanimous.[45]

Ian Kennedy, in analyzing one of the early neonatal withdrawal of treatment decisions, is even more explicit about how the best interests test is effectively equated with medical opinion. Commenting on Sir Stephen Brown's judgment in Re C (1992) he contends:

> The court has to decide whether this or that is in a child's best interests. We all know that in medical law cases this is at best a face-saving device, allowing the court to endorse the view of doctors while appearing to be in charge. But at least the proprieties are observed.[46]

Thus, in most cases, the "best interests" test is reduced to a device for rubber stamping medical decisions and discounting the opposing view of parents, although there are recent indications that, in line with a greater willingness throughout healthcare law to hold doctors to account,[47] that the test is being used to allow greater scrutiny of professional reasoning. Yet, in the case of male circumcision, there is an interesting reversal of the way in which the best interest test generally operates. Far from discounting parental choices, it serves in this instance to shield them from medical and judicial scrutiny. In the *Re J* case, Wall J. is categorical that clinical considerations are not decisive, since the assessment of 'best interests' clearly includes religious, cultural and social benefits. To some extent, this aspect of his judgment is in line with other recent cases. Hence, in the context of an application to sterilize a mentally incompetent woman, Thorpe J (who sat in the Court of Appeal in *Re J*) had earlier noted that considerations of patient welfare "embrace issues far wider than the medical. Indeed, it would be undesirable and probably impossible to set bounds to what is relevant to a welfare determination."[48] Similarly, in issuing a declaration that authorized the use of untested experimental treatment on incompetent patients with CJD, Dame Butler Sloss stated that her task was:

> to assess the best interests in the widest possible way to include the medical and non medical benefits and disadvantages, the broader welfare issue of the two patients, their abilities, their future with or without the treatment, the view of the family and the impact of refusal of the applications.[49]

However, notwithstanding attempts in these more recent judgments to construct something akin to a checklist of best interests which renders the cost–benefit equation more transparent, it remains true that the policy factors that underlie such tests "are rarely fully articulated, much less satisfactorily resolved."[50] Nor has there been any judicial indication of how clear and immediate medical risks could adequately be balanced against putative social benefits to accrue in the future. Rather, as we have sought to demonstrate, in the context of infant male circumcision, a shared medico-legal common sense of the practice as uncontroversial and mainstream militates against an exploration of the risks and harms of the practice. As John Harrington has argued in another context, the "best interests" test can never function as a rule "with a fixed, univocal meaning which need only be applied to the facts of the case at hand."[51] Rather, he contends that the test functions as a guiding standard, significantly influenced by the "intuitive sense of reasonableness of the deciding judge":

> [G]uiding standards open up the system by allowing judges, under the influence of changing social mores and scientific developments, to fashion and refashion new criteria of best interests etc.... The courts respond to what they see as change by working and re-working a distinctively judicial common sense. Values and principles, but also stock images and stereotypes are the stuff of this common sense.[52]

In the male circumcision context, we would argue that this judicial common sense incorporates a notion of male circumcision as a routine and benign surgical procedure performed by caring and well-intentioned families — precisely the sort of common sense that informs the professional guidance governing the practice. Ellen Feder points out that to query such practices "demands that we investigate areas of understanding resistant to critical examination — areas we call 'common sense' or 'what goes without saying.' "[53] Yet, as we have traced in earlier papers, in reality, when subjected to critical scrutiny, it is readily apparent that circumcision has long existed as a procedure in search of a rationale. At different times, circumcision has been promoted as a remedy for alcoholism, epilepsy, asthma, curvature of the spine, paralysis, malnutrition, night terrors, clubfoot, eczema, convulsions, promiscuity, syphilis, and cancer.[54] All of these justifications have been debunked or at best proven inconclusive, thereby confirming Hilde Lindemann's observation that "the social norms that inevitably form some part of a parental conception of what life is about are not the product of rational individual reflection and personal choice."[55]

Hence, the problem with invoking such an elastic and subjective test as "best interests" as a guide lies in its inability to contest practices that have become normalized by a social system in which circumcision is validated. The extent to which this has occurred is revealed in the following quotation from Caroline Bridge:

> Male circumcision is relatively harmless albeit, like tattooing, irreversible. The risks are minimal, the cultural and religious significance very great to large number of people, and it is now reported as having long term protective effects. This enhances its rational basis. Just as significantly, *it is not a practice belonging to the fringes of society but is almost part of the mainstream*. Female circumcision is the reverse.[56]

We would suggest that it is precisely due to this status as "almost part of the mainstream," that male circumcision, unlike other practices, including female genital cutting or scarification, is normalized, rather than being cast as a religious or cultural practice which requires justification. In similar vein to our contention here, Ellen Feder has argued, in the context of genital surgery on children who are born with intersex conditions, that medics and parents tend to follow certain 'rules of normality.' She contends:

> [T]hese rules are not the rules of mere social convention, but something more along the lines of what might be described as 'cultural unconscious,' conventions that are not considered and weighted, thoughtfully enacted *by* individuals but conventions that could more precisely be understood to work *through* individuals.[57]

This observation helps explain why an often unreflective family preference for these forms of genital cutting is readily deemed to fall within the range of legally acceptable parental choices, which allow other interests, including the bodily integrity of a child too young to make his own choice, to be over-ridden. The fact that these choices are endorsed or promoted by certain medical practitioners serves to further shield them from judicial scrutiny.

Our related concern with the judgment in Re J and the BMA guidance pertains to the assumption that parents are best placed to articulate and defend the interests of children. This view has been challenged by O'Donovan and Gilbar, who have argued (in the context of decision-making on behalf of incompetent adults), that "[f]amilies and significant others are ambivalently placed."[58] Such ambivalence, they argue, stems from a tension between, on the one hand, seeing the patient as an individual and, on the other, recognizing the reality of interdependence and the possibility for conflicts to which this may give rise. In the case of the young child, his dependence on his parent makes it harder still to disentangle the interest of parent and child.[59] Feder argues that studies involving the decision whether to consent to genital cutting for children born with intersex conditions,[60] have pointed to "a contradiction between what individuals would want for themselves and what they would feel is right for their children."[61] She proposes that parents in this situation should seek to *identify with* the child rather than to consider *what is right or best for them*.[62] Similarly, we would advocate that judges attempt something akin to such a substituted judgment test, given that some past cases have exhibited an over-identification on the part of judges with the situation of the parent, rather than the child, and have led to courts endorsing parental choices that appear to lack reason.[63] In the context of circumcision, the problem of simply condoning the choices parents may make for their children is compounded by ample evidence that many parents simply do not understand to what they have consented. Hence, one study found that 87% of mothers interviewed were not aware of the risks of circumcision.[64] Moreover, in assessing the rationality of the decision — a factor which we regard as crucial in applying the "best interests' test[65] — another study found that the most widely cited reason for the parents' choice to circumcise (46%) was a desire for their son to resemble other males.[66] While not wishing to impugn the benign intentions of parents who choose to circumcise their children, we would follow Elliston in arguing that decisions taken by parents on behalf of children too young to consent for themselves should be required to meet the standard of "ensuring that significant interests of the child are not put at risk and that the decisions made by parents meet a reasonableness standard."[67] Decisions taken in the absence of salient information about risks, or rooted in a desire for the child to resemble its parent in this respect, would appear not to satisfy such a standard.

Alternatives to the "Best Interests" Test

In our view, attempts to redefine the "best interests" test are unlikely at present to prompt different outcomes in cases involving the circumcision of male children, so we would argue instead for a different way of framing the issue. John Eekelaar has cautioned that: "The 'welfare' or 'best interests' principle, from time to time, has been subjected to critical scrutiny. But it is easier to criticize the principle than to come up with an alternative."[68] Nevertheless, while conceding the difficulties in framing alternatives, Elliston argues that the time has come to replace the best interests standard with a "simpler and more coherent approach [which]...would... require significant risk of serious harm as the threshold for judicial intervention" in addition to subjecting parental views to a reasonableness criterion.[69] We share her view that the notion of harm is key and that parents must be required by law to justify any decision which inflicts harm. However, we would add that a further shift in discourse may be productive in order to highlight or uncover the harms inherent in practices that have become normalized, and hence invisible, in our medico-legal culture.

Interestingly, there are hints in a later circumcision judgment of just such an alternative discourse to that of "best interests." In Re S, an application to circumcise a 9-year-old boy and to convert he and his sister to the Muslim faith was made by the children's mother who had separated from their father — a member of the Jain faith.[70] Unsurprisingly, given his earlier leading judgment in Re J, a key element of Thorpe LJ's short judgment, upholding Baron J's refusal to grant the declaration, was that these children had a mixed Muslim/Jain cultural heritage. However, surprisingly the judgment contains no reference to the "best interests" test. We argue that this and other aspects of the judgment can be interpreted as suggestive of a new approach, although we would concede that they are not articulated fully. First, in this case, there is an explicit attempt to disentangle the interests of the parents from those of the child. Thorpe LJ notes that the mother was motivated more by her desire to cement her relationship with a Muslim partner than by concern for her children's welfare, thus acknowledging the potential for parents to use claims about child welfare to advance their own interests. Secondly, we would suggest that there is potential in the way that Thorpe LJ frames the scope for such conflict:

> [T]he current problem stems not from the children's *needs* but from the need of the mother to portray her marriage as being to a Muslim man.[71]

The language of "needs," we argue, lends itself more readily to opportunities for the judge to separate out the different and possibly competing interests or rights that are at stake. Thus, framing the issue as a question of how best to meet the "needs" of the child, rather than the nebulous notion of promoting his "best interests," may offer a productive way to re-think decisions about circumcision. In our view, the discourse of "needs" helps to foster the identification with the child's position advocated by Feder.[72] Moreover, in terms of satisfying the court that the relevant threshold for decision-making has been met, we contend that it would be considerably harder to construct a reasoned argument that any young child *needs* to be circumcised (even in a cohesive family characterised by religious observance) than it is to satisfy a test that requires merely that performing such surgery is *not contrary* to that child's interests. The third positive feature of the *Re S* case is that Thorpe LJ upholds Baron J's pragmatic ruling that the son should be permitted to make his own informed decision as to whether to undergo circumcision once he has attained *Gillick*-competence.[73] We see this solution as one that accords with recent trends in the treatment of children born with intersex conditions,[74] and which is gaining acceptance as the most appropriate approach in the case of male circumcision.[75] Indeed, this approach has also emerged in communities where circumcision is part of religious practice.[76] This emerging consensus about postponing non therapeutic surgeries until the child has competence to decide, has, as we have argued elsewhere, been the product of a movement to uncover the harms inflicted by early non-consensual intersex surgery. In the case of male circumcision, clearly much work remains to be done on this process of uncovering harms, given that so many of the harms of the practice are rendered invisible even to "caring" or "good" parents.[77]

Thus, we suggest that, in addition to the medical risks briefly outlined above, it is also necessary for decision-makers to take into account issues of bodily integrity, the desirability of keeping future choices open, and the possibility of psycho-sexual harms as well as possible negative effects on future sexual experience and enjoyment. Such issues can be brought to the fore if we pay attention to how some men experience their circumcision status. Qualitative studies, such as that conducted by Hammond in the US and published in the *British Journal of Urology International* in 1999, do much to complicate the idea that male circumcision is sufficiently de minimus that it should be left to parental choice.[78] Hammond's study ably details the range of negative physical, sexual, and emotional effects that may follow routine juvenile circumcision. While acknowledging the particularity of such experiences, simultaneously we need to recognize that a general failure to unpack these harms may be attributed not only to the unwillingness of doctors and parents to see them, but also to deeper-rooted problems with the concept of harm. Indeed, part of the problem is that, as Joanne Conaghan points out, harm "is widely assumed to be self-evident."[79] We would argue, following Robin West's analysis of harm to women, that law consistently ignores, legitimizes, and minimizes harms sustained by children. As she notes, because of legal culture's legitimizing power, harms that it fails to recognize effectively disappear.[80] For this reason, although we find much of the critical discourse around female circumcision and the punitive legal response to it problematic, we would contend that one positive feature of how female circumcision is legally regulated is the unambiguous acceptance that the procedure is harmful. We would certainly argue that all forms of harm inflicted on young children whose bodies are moulded and redesigned by surgeons are comparable, regardless of whether the motivation is to "normalise" or "perfect." The harms of male circumcision, however, have been rendered less visible and contentious by the long history and widespread acceptance of the practice in North America, the United Kingdom, and Australia.

Conclusions

In this essay, we have traced a number of objections to the position that parental choice is the key determinant of non-therapeutic neonatal male circumcision, challenging in particular the efficacy of the amorphous "best interests" standard in protecting children from harm. And "harm" is where we would like to refocus discussion. What we argue for is an analysis grounded in the concepts of children's needs and the prevention of harm. Whilst it may well be that discussion of needs and harms collapses to become indistinguishable from current "best interests" calculations, we would contend that altering the legal standard may — even if only briefly — challenge us to reflect more critically on our judgments of what is right for children. Requiring that we articulate our arguments, justifications and concerns within a new paradigm which foregrounds harm does have the potential (however slight) to shift our current understanding and practices, and problematise common sense assumptions. Furthermore, we would suggest that coupling the notion of "harm" with the concept of "need" affords some protection against the risk of a new harm-based test disintegrating into something indistinguishable from the current "best interests" calculation. Clearly adopting a framework grounded in harm alone is double-edged. Michael Freeman, for instance, has used the language of harm to argue that "the failure by Jewish or Muslim parents to circumcise their sons is a form of abuse, with the child being likely to suffer 'significant harm.'"[81] We would contend, however, that Freeman's reasoning implicitly pertains to possible future harm and, while there may be legitimate concerns as to matters of familial, social, and religious identity and belonging in his future, it is difficult to extrapolate from that position to one that the child *needs* to be circumcised as a neonate.

Thus, in our view, a stress on the child's immediate needs, coupled with alertness to the possibility of tangible present harm, mandates an approach to routine neonatal circumcision that is better aligned with accepted medico-legal standards for the care of children and protection of their rights. If the concern is with future harms of cultural identification and belonging, then a symbolic act followed by elective circumcision once the child is *Gillick*-competent (and hence capable of weighing up the health and cultural impacts) is, we would argue, a more appropriate response. In addition to responding to cultural harms that may well be better addressed once the child can make an informed choice, we would argue that focusing on needs foregrounds the more immediate harms and risks of the procedure. On this view, the risks of limiting sexual experience, causing harm to sexual functioning, and creating further adverse outcomes, are more legitimate concerns for doctors charged with assessing parental requests for circumcision, than broader social judgments which they are ill equipped to make. So long as social judgments are given space and credence in law, it is clear that infant male circumcision will continue to be constructed in professional codes and by legal norms as a risky practice, rather than a harmful one, and thus as a private matter to be appropriately decided by parents.[82] However, we regard it as both significant and welcome that our preferred discourse of present harm and needs is beginning to enter ethico-legal debates about the appropriateness of circumcision.[83, 84] We suggest that a process of dialogue and education, rather than calling for criminal prohibitions and sanctions, is the appropriate strategy to counter male circumcisions and our common sense assumptions that it is a trivial procedure. In addition to, and as part of, this process of changing parental attitudes education must be targeted at healthcare providers and the professional bodies that promulgate guidance on this issue. However, in the meantime, civil courts faced with a request to authorize circumcision on a child too young to consent should hold parents to a standard of decision-making that is informed about harms, sensitive to needs, and demonstrably reasonable.

References

1. Earlier versions of this paper have been presented at the 9th International Symposium on Genital Integrity and Human Rights 24–26 August 2006, University of Washington, Seattle; at the British Academy and Feminism and Legal Theory Workshop ‘Exploring Key Concepts in Feminist Legal Theory: The Family,’ Emory University, Atlanta, September 7–8, 2007, and at the AHRC Centre LGS Workshop on Body Modification, University of Keele, 24 October 2007. We would like to thank the participants at these events for their feedback which has helped shape this paper.
2. See, for instance, J. Eekelaar, “The Emergence of Children’s Rights” (1986) 6 Oxford Journal of Legal Studies 161; Report of the Inquiry into Child Abuse in Cleveland, 1987, Cm 412 1988; Children Act 1989, M. Freeman, “Taking Children’s Rights More Seriously” (1992) International Journal of Law and the Family 52.
3. A. Bainham, “Children Law at the Millenium” in S. Cretney (ed) Family Law: Essays of the New Millenium, London: Family Law, 2000 at p. 11; see also J. Bridgeman and D. Monk, “Introduction” in Bridgeman and Monk (eds) Feminist Perspectives on Child Law, London: Cavendish, 2000.
4. Prohibition of Female Circumcision Act 1985, Female Genital Mutilation Act 2003.
5. W.E. Brigman, “Circumcision as Child Abuse: The Legal and Constitutional Issues” (1984–1985) 23 Journal of Family Law 337.
6. Other developed Western European nations have a cumulative national circumcision rate of only c 1.6% for boys under 15 years of age — see J.P. Warren, “NORM UK and the Medical Case Against Circumcision: A British Perspective” pp. 85–101 in Denniston and Milos (eds) Sexual Mutilations: A Human Tragedy, Plenum Press: New York, 1997, at 86; M. Frisch et al., “Falling incidence of penis cancer in an uncircumcised population (Denmark 1943–1990)” BMJ 311 (1995) 1471; S. Waldeck, “On the Cutting Edge: Social Norm Theory and Male Circumcision,” paper given at an ‘Uncomfortable Conversation’ — Sexuality and Feminist Theory: Road Blocks, Detours and New Directions (15th–16th November 2002, Cornell Law School), 1 at 9 (copy on file with the authors); see also, Queensland Law Reform Commission, Circumcision of Male Infants (Brisbane: QLRC, 1993).
7. D.L. Gollaher, “From Ritual to Science: The Medical Transformation of Circumcision in America” (1994) Journal of Social History 5 at 25. See also, D. Gairdner, “The Fate of the Foreskin: A Study of Circumcision” II (1949) BMJ 1433–1437.
8. A.M.K. Rickwood and J. Walker, “Is Phimosis Over Diagnosed in Boys and Are Too Many Circumcisions Performed in Consequence?” (1989) 71 Annals of the Royal College of Surgeons of England 275.
9. Gairdner, op. cit., note 7.
10. Rickwood and Walker, op. cit., note 8.
11. Editorial, “The Case against Circumcision” (1979) 6127 BMJ 1163–1164.

12. A.M. Johnson, J. Wadsworth, K. Wellings, J. Field, and S. Bradshaw, Sexual Attitudes and Lifestyles, Blackwell Scientific Press, Oxford, 1994.
13. N. Williams and L. Kaplila, "Why are Children Referred for Circumcision?" (1993) 306 BMJ 28.
14. L. Rangecroft, "A Rite to Drop" (1995) 311 BMJ 816.
15. Warren op.cit., note 6 at 86.
16. A.M.K. Rickwood, S.E. Kenny, and S.C. Donnell, "Towards Evidence Based Circumcision in English Boys: Survey of Trends in Practice" (2000) 321 BMJ 792.
17. Ibid.
18. Ibid.
19. S.S. Dave, A.M. Johnson, K.A. Fenton et al. "Male Circumcision in Britain: Findings from the National Probability Sample Survey" (2003) 79 Sexually Transmitted Infections 499–500.
20. P. Cathcart et al. "Trends in Paediatric Circumcision and Its Complications in England between 1997 and 2003" (2006) 93 British Journal of Surgery 885–890.
21. M. Fox and M. Thomson, "Short Changed? The Law and Ethics of Male Circumcision" (2005) 13 International Journal of Children's Rights 161–181.
22. For a discussion of circumcision and medical responses to intersex conditions see M. Fox and M. Thomson, "Cutting it: Surgical interventions and the sexing of children" (2006) 13 Cardozo Journal of Law and Gender 101–117.
23. M. Fox and M. Thomson, "A Covenant with the Status Quo: The New BMA Guidance on Male Circumcision" (2005) 31 Journal of Medical Ethics 463–469.
24. Available at http://www.bma.org.uk/ap.nsf/Content/malecircumcision 2006.
25. M. Benatar and D. Benatar, "Between Prophylaxis and Child Abuse: The Ethics of Neonatal Male Circumcision" (2003) 3 American Journal of Bioethics, 35–48, at 42.
26. R. S. Van Howe, "A Cost-Utility Analysis of Neonatal Circumcision" (2004) 24 Medical Decision Making 584–601.
27. British Medical Association, The Law and Ethics of Male Circumcision: Guidance for Doctors (London: BMA, 2003). References are to the 2003 guidelines unless otherwise stated. The most recent guidelines are available at http://www.bma.org.uk/ap.nsf/Content/malecircumcision 2006.
28. Ibid at 7.
29. Ibid at 3. Note the similarities to the position defended by Benatar and Benatar, op. cit. at note 25.
30. http://www.bma.org.uk/ap.nsf/Content/malecircumcision2006.
31. Ibid.
32. Certainly in comparison to the guidance issued by the American Academy of Paediatrics, "Task Force on Circumcision, Circumcision Policy Statement" (1999) 103 Paediatrics 686–693 — see R. Darby and J. S. Svoboda, "A Rose by Any Other Name? Rethinking the Similarities and Differences between Male and Female Genital Cutting" (2007) 21 Medical Anthropology Quarterly 301–323 at pp 313–315.
33. See Benatar and Benatar, op. cit., note 25, at pp. 37–38 and accompanying references; F. Warnock and D. Sandrin, "Comprehensive Description of Newborn Distress Behavior in Response to Acute Pain (Newborn Male Circumcision)" (2004) 107 Pain 242–255.
34. N. Williams and L. Kapila, "Complications of Circumcision" (1993) 80 British Journal of Surgery 1231–1236; E.W. Gerharz and C. Haarmann, "The first cut is the deepest? Medicolegal aspects of male circumcision" (2000) 86 BJU International 332–338.
35. F.M. Hodges, J.S. Svoboda, and R.S. Van Howe, "Prophylactic Interventions on Children: Balancing Human Rights with Public Health" (2002) 28 Journal of Medical Ethics 10. For a recent reported instance of death as a result of circumcision in the UK see S. Moyes, "7-Day-Old Died after Circumcision" Daily Mirror 15 February 2007.
36. Re J (A Minor) (Prohibited Steps Order: Circumcision), sub nom Re J (Child's Religious Upbringing and Circumcision) and Re J (Specific Issue Orders: Muslim Upbringing & Circumcision) (2000) 1 FLR 571; (2000) 1 FCR 307; (2000) 52 BMLR 82.

37. See M. Fox and J. McHale, "In Whose Best Interests?" (1997) 60 Modern Law Review 700; A. Grubb, "Treatment Decisions: Keeping It in the Family" in A. Grubb (ed) Choices and Decisions in Health Care, Chichester: Wiley, 1993.
38. S. Colb, "Divorce, Religion, and Circumcision: What a Conflict Tells Us about Parental Rights" Find law: Legal News and Commentary, 28 November 2007.
39. BMA, Consent, Rights and Choices in Health Care for Children and Young People, London: BMJ Books, 2001.
40. See, in particular, the judgment of Thorpe J in Re A (medical treatment: male sterilisation) (2000) 1 FLR 549.
41. For a recent analysis see S. Elliston, The Best Interests of the Child in Healthcare, London: Routledge-Cavendish, 2007.
42. Thus, for instance, the Medicines for Human Use (Clinical Trials) Regulations 2004 (which implement the EU Clinical Trials Directive in the UK) require that the child's parent or legal representative should act on the basis of the minor's 'presumed will' (see Article 4(a) of the Clinical Trials Directive). In this context a 'substituted judgment' test effectively replaces the common law test of best interests in considering the enrolment of a minor under 16 in clinical research — see M. Fox, "Clinical Research and Patients: the Legal Perspective" in Tingle and Cribb (eds) Nursing Law and Ethics, 3rd edition, Blackwells, 2007.
43. See M. Brazier, "Letting Charlotte Die" (2004) 30 Journal of Medical Ethics 519–520; M. Brazier, "An Intractable Dispute: When Parents and Professionals Disagree" (2005) 13 Medical Law Review 412.
44. See, for instance, An NHS Trust v. MB (2006) EWHC 507; R (Burke) v. GMC (2005) EWCA 1003.
45. S. Pattinson, Medical Law and Ethics, London, Sweet & Maxwell, 2006 at p. 506.
46. I. Kennedy, "Child: Discontinuation of Treatment"; Re C (A Baby) (1997) 5 Medical Law Review 102–104 at p. 103.
47. See, for instance, M. Brazier and J. Miola, "Bye Bye Bolam" (2000) 8 Medical Law Review 84–115.
48. Re S (Adult Patient Sterilisation) 2001 Fam 15; see also Re A (medical treatment: male sterilization) (2000) 1 FLR 549; A v A Health Authority (2002) 1 FCR 481; R on the application of N v. Doctor M and others (2003) 1 FLR 667.
49. Simms v. Simms and another (2003) 1 All ER 699.
50. Fox and McHale, op. cit. note 37 at p. 701.
51. J. A. Harrington, "Deciding Best Interests: Medical Progress, Clinical Judgment and the 'Good Family'" (2003) 3 Web JCLI.
52. Christine Piper has contended that in Child Law generally "In relation to establishing the best interests of children there have emerged strong general presumptions which constrain the process of working out what is best for individual children in the situation in which they find themselves" — C. Piper "Assumptions about Children's Best Interests" (2000) 22 Journal of Social Welfare and Family Law 261 at p. 262.
53. E.K. Feder, "'In Their Best Interests': Parents Experience of Atypical Genitalia" in E. Parens (ed) Surgically Shaping Children: Technology, Ethics, and the Pursuit of Normality, Baltimore: John Hopkins University Press, pp. 188–210.
54. G.P. Miller, "Circumcision: Cultural-Legal Analysis", (2002) 9 Virginia Journal of Social Policy and the Law 497 at pp. 502–503.
55. H. Lindemann, "The Power of Parents and the Agency of Children" pp. 176–188 in Parens op. cit. note 53 at p. 179.
56. C. Bridge, "Religion, Culture and the Body of the Child" in A. Bainham et al. (eds) Body Lore and Laws, London, Hart Publishing 2002 at pp. 265–287 at p. 284 (our emphasis).
57. Feder, op. cit. note 53 at 192.
58. K. O'Donovan and R. Gilbar. "The Loved Ones: Families, Intimates and Patient Autonomy" (2003) 22 Legal Studies 332 at p. 333.
59. Indeed as Hildeman suggests, op. cit. note 55 at p. 182, it is arguably undesirable for caring parents to "mark a clear boundary between themselves and their children. In a properly loving

relationship that boundary ought to be blurred in at least as many places as it remains distinct."

60. See S. Kessler, Lessons from the Intersexed, New Brunswick NJ: Rutgers University Press, 1998.
61. Feder, op. cit. note 53 at p. 206.
62. Ibid.
63. See the judgment of Butler Sloss LJ in Re T (a minor) (wardship: Medical treatment) (1997) 1 WLR 252 at p. 251: "This mother and child are one for the purpose of this unusual case and the decision of the court to consent to the operation jointly affects the mother and son and vit also affects the father. The welfare of this child depends upon his mother."
64. See J.R. Wayland & P.G. Higgins, "Neonatal Circumcision: A Teaching Plan to Better Inform Parents" (1982) 7 Nurse Practitioner 26.
65. See Fox and McHale, op. cit. note 37; Elliston op. cit. note 41.
66. A. Henerey "Evolution of Male Circumcision as Normative Control" (2004) 12 Journal of Men's Studies 265.
67. Elliston, op. cit. note 41 at p. 2.
68. J. Eekelaar, "Beyond the Welfare Principle" (2002) 14 Child and Family Law Quarterly 237 at p. 237. See also Elliston, ibid at p. 287.
69. Elliston,op. cit. note 41 at pp. 2–3.
70. Re S (children) (Specific Issue: Religion: Circumcision) (2005) 1 FLR 236.
71. Ibid at p. 238 (our emphasis).
72. See note 53 above.
73. The landmark House of Lords ruling in Gillick v. West Norfolk and Wisbeck Area Health Authority (1985) 3 All ER 402 held that parental rights to decide should yield to the rights of a child who possessed sufficient maturity and understanding to appreciate the implications of the decision.
74. A. Domurant Dreger, "What to expect when you have the child you weren't expecting," pp. 523–566 in Parens op. cit. note 53 at p. 259.
75. J. Fortin, Children's Rights and the Developing Law, 2nd edition, London: Butterworths, 2005 at pp. 329–332, Ellison, op. cit. note 41 at pp. 98–99; G. Hinchley, "Is Infant Male Circumcision an Abuse of the Rights of the Child?" (2007) 335 BMJ 1180.
76. Hinchley ibid.
77. See Harrington, op. cit. note 51.
78. Darby and Svoboda have argued that the "most obvious and universally experienced harm of all [is] the deprivation off an integral, visually prominent, and erotically significant feature of the penis." op. cit. note 32 at p. 304.
79. T. Hammond, "A preliminary poll of men circumcised in infancy or childhood" (1999) 83 (supp 1) BJU International 85. See also, R. Darby and L. Cox, "Objections of a sentimental character: The subjective dimension of foreskin loss" (2007) unpublished paper (copy on file with authors).
80. J. Conaghan, "Law, Harm and Redress: A Feminist Perspective" (2002) 22 Legal Studies 319–339.
81. R. West, Caring for Justice, New York: New York University Press, 1997.
82. M. Freeman, "A Child's Right to Circumcision" (1999) 83 BJU Int 74–78.
83. See Fox and Thomson, op. cit. note 21.
84. See, for instance, the British Association of Paediatric Surgeons; Statement of the Working Party on the Surgical Management of Children Born with Ambiguous Genitalia, London, 2001; British Medical Association, The Law and Ethics of Male Circumcision: Guidance for Doctors London, 2006.

Cultural Relativism at Home and Abroad

An American Anthropologist Confronts the Genital Modification of Children

Zachary Androus, BA, MA, and PhD Candidate in Race, Gender and Social Justice

Abstract Cultural relativism acknowledges the meaningfulness of cultural behaviors for those who practice them, and is a central tenet of anthropology. In cases of apparent conflict between culture practices and universal human rights, anthropologists articulate sophisticated positions that move beyond the simple dualisms inherent in absolutist condemnations or defenses of controversial behaviors. Approaching cultural practices that are contrary to the researcher's own values does not require the suspension of the researcher's moral or ethical sensibilities; however, it does require the researcher to remain sensitive to the values of the people with whom they work, especially when dealing with contested practices like circumcision. Anthropologists who encountered female genital modification have written extensively on their application of cultural relativism, In my work on American neonatal circumcision, I consciously apply the principles of cultural relativism, despite my personal opposition to the practice. This paper describes the importance of a relativist and considers its implications for intactivism in an American context.

As a cultural anthropologist, I am, in the broadest terms, a social scientist who studies human behavior. As a medical anthropologist, I am specifically concerned with human behavior related to concepts and systems of health and healing and also with medical systems. Concepts of health and medical systems are commonly related, but, in some cases, they also operate independently of one another. All such systems manifest in ways that involve some form of bodily manipulation, as most broadly conceived. If we consider the broad collective category of human genital modification, a good deal of behavior around the world included in that category falls within the purview of medical anthropology. My doctoral research, which is currently underway (as of Autumn 2006), focuses on how different sources of information influence parental decision making for or against neonatal male circumcision in the United States and, because I am an anthropologist, my research methodology is grounded in a strategy called cultural relativism. It is important at

Department of Anthropology, American University, Lorenzo de Medici University, Florence, Italy

G. Denniston et al. (eds.), *Circumcision and Human Rights*,

the outset that I make clear that I am personally opposed to the practice of neonatal male circumcision. I grew up with a cultural bias against circumcision (my mother and my father's father both came to the United States from Greece, where circumcision is not generally practiced, and I remember when I was a little boy my father used the word "barbaric" to characterize the American practice of circumcision when explaining to me the difference between my penis and those of my playmates. I am fascinated by the apparent persistence of Hellenic values from the ancient past represented by his use of that particular word, descended into English from the Greek word for foreign; see Hodges (2001)[1] for more on Hellenic genital aesthetics), and as an adult I find the bioethical arguments against elective surgery on infants by proxy consent to be very compelling. But my personal opposition to the practice is not the basis from which I approach research into the practice, and I am very careful not to challenge the choices for or against circumcision made by the parents with whom I conduct my research.

In this paper, I discuss the influence of the anthropological approach to female genital modification, which is characterized by cultural relativism, on my approach to research on male circumcision in the United States. These issues are fundamentally different: not because of anything to do with male genital modifications versus female genital modifications, but because African nations and the United States present fundamentally different social, cultural, and political contexts. This is addressed in further depth below, but first I consider what cultural relativism is and what it is not. I then touch on the anthropological approach to female genital modification, and then I explain how all of this informs my own approach to studying circumcision in the United States. Being a paper on cultural relativism, this is really a paper about methodology. This is my narrative on how I reconcile my personal beliefs with the methodological protocols that have been developed to help effect quality social science research.

Cultural relativists frequently find themselves doing what I am about to do now, which is immediately set out to explain how cultural relativism is not the same as moral relativism or ethical relativism. Moral and ethical relativism are names for ideologies that hold that, because there are no moral absolutes, everyone should be free to observe their own individual moral or ethical standards. Moral relativism is kind of a political-philosophical straw man; many people declaim the threat of moral relativism but, according to the *Stanford Encyclopedia of Philosophy*, very few people actually promote it in a serious way.[2] Cultural relativism is a perspective that recognizes the fact that people behave in ways that are meaningful to them based on their own cultural values and beliefs. This does not mean their own individual values and beliefs, this means values and beliefs shared within particular societies. Each society has a moral and ethical code: a system for determining right and wrong that is institutionalized and enforced to some greater or lesser extent. I am aware of no society on record in which what is moral or ethical behavior is determined at the individual level. Cultural relativists do not make the claim that because different moral and ethical standards exist across different societies that there should be no such thing as a moral standard. Cultural relativism is really not a claim to that kind of absolute philosophical knowledge of the moral order of the universe;

what cultural relativism is really is a perspective: a way of looking at human behavior, a strategy for thinking about human behavior. The essence of cultural relativism is the idea that, when we encounter human behavior that strikes us as strange, unusual, inappropriate, or even morally or ethically repugnant, we have to look at it in terms of the local standards of conduct. How does such behavior measure up to the standards of the society whose members are practicing it? In every society, moral and ethical standards are continually changing through time, so you may not find consensus on what is right and wrong, but you will always find some system of determining right and wrong behavior in every society.

So cultural relativism is absolutely not making the claim that there is no right and wrong, or that there shouldn't be moral or ethical standards. Instead, cultural relativism is an attempt to account for and deal with the fact that there are multiple systems of right and wrong operating in the world, each with different standards that apply in their own particular contexts. Anthropologists developed cultural relativism out of their experience that effective social scientific inquiry into one society cannot be based on the belief system of another society. An approach that uses a single set of moral and ethical positions as a standard against which to measure the acceptability of other such sets of ideas is problematic in part because most groups of people in the world tend to hold their own system of right and wrong as a standard against which to measure others. Ironically, this tendency is displayed by the moral absolutists who build the straw man of moral relativism and promote the idea of conflict between those philosophies that promote a universal moral standard in the world and those that recognize a diversity of moral standards in the world.[3, 4]

This apparent conflict plays out when people who hold universal moral standards encounter other people whose behavior is unacceptable to that standard. The world has been globalized for hundreds of years; what happens when people from one society with its own moral code, encounters another society that practices some strange and terrible custom that offends said moral code? Well, conflict results, and these kinds of conflicts are not easily resolved. The issues associated with them are not necessarily difficult to understand, but they can be extremely challenging to engage.

A widely cited example of cultural relativism is the anthropological response to female genital modification, or female circumcision. Since the 1970s, female genital modification, or FGM, has been a topic of heated debate in the fields of international public health and human rights. Just the debate over what to call these practices, and whether the "M" in FGM should stand for the value-neutral term "modification" or the value-negative term "mutilation," has been contentious.[5,6]

In the context of the FGM debates, cultural relativism is a frequent target of criticism from scholars, activists, and policymakers; for example, medical anthropologist Melvin Konner used a review of Hanny Lightfoot-Klein's book, *Prisoners of Ritual*, as an opportunity to criticize relativist ethnographers as "soft" on female circumcision, dismissing Lightfoot-Klein's attempt to account for the insider's perspective "because she has come to sympathize so well with the folk view."[7] Konner goes on to opine that "cultural relativism has limits, and this [female circumcision] is one place where we ought to draw the line."[7] Konner's "draw the line" comment was cited by Daniel Gordon (1991)[8] in a prominent anthropology journal, who was in turn met with scathing rejoinders about his role in the "Western civilizational project" (Morsey 1991:19) undertaken as part of the neo-colonial endeavor of "authoritative Western humanism" (Morsy 1991)[22]. These debates are ongoing: recently Gerry Mackie (2003)[9] published a mildly combative critique of Carla Makhlouf Obermeyer (1999),[10] who then replied with a reaffirmation of her earlier positions (2003).[10]

Nevertheless, anthropologists who work with groups of people who practice female genital modification tend to be reticent about issuing blanket condemnations of the practice, even when they are careful to point out that they do not support the practice (e.g., Boddy 1991[23]; Sargent 1991[24]; Gruenbaum 2001:20–24;198–202[5]). So why not "draw the line" on cultural relativism when it comes to female circumcision? There are philosophical and ideological reasons, and there also are practical, technical reasons. Putting aside the philosophical debate for a moment, and simply looking at conditions on the ground, so to speak, isolating female circumcision as an issue unto itself is a very poor way to address the overall quality of life for the people whose genitals the foreigners want to protect. A Somali immigrant in Italy remarked to an Italian healthcare worker that Europeans seemed to care more about the clitorises of African women than they cared about the women themselves (Noel Gazzano, personal communication, April 20, 2006). There is a pragmatic dimension to the role of cultural relativism in anthropological practice; according to medical anthropologists George M. Foster and Barbara Gallatin Anderson:

> [T]he anthropologist's emphasis on cultural relativism is not simply a broad minded plea for tolerance of the ways of others; it is an essential foundation for successful technical aid, in health and in all other fields. The operational rule underlying the principle of cultural relativism is that before attempting to implement change, one must learn the reasons why the traits under attack are present, the roles they fulfill, and their meanings to the people.[11]

From an operational perspective, applied social scientists see that interventions aimed at health-related behaviors are ineffective otherwise, and making claims to absolute moral and ethical knowledge in the face of practices that are acceptable in their own social contexts is not the best strategy for engagement with the people whose practices are perceived to be problematic by outsiders.

Data from a recent World Health Report indicates that communicable diseases are the primary health concern facing Africans (Semakula 2002:1).[12] Poverty is the primary underlying condition contributing to early mortality in Africa, and more children suffer the consequences of being underweight and lacking access to clean water and adequate sanitation than suffer complications from culturally motivated surgeries. If all female genital operations in Africa were to stop today, thousands of children would still die tomorrow from diarrhea brought on by communicable diseases that result from lack of access to sanitation and clean water, conditions which are in turn directly related to poverty. In the populations that practice female circumcision, the overall public health risk from circumcision is lower than it is from malaria or tuberculosis (Semakula 2002:1–3).[12] Limited access to clean water and adequate sewage, insufficient access to healthcare systems that are inadequate even when they are accessible, and in some cases even armed conflict,[12] all present greater threats to public health in Africa than does circumcision.

The point of this is to say that, when it comes to identifying public health issues, populations are best served by both a holistic approach and what can be thought of as a sort of public-health triage. I personally find it difficult to justify a focus on anything without also committing resources to address the lack of clean water and sanitation. When faced with these other issues, the singling out of genital modification, and the disregard for its local significance, has created confusion and hostility in many African communities (Gruenbaum 2001:203–205).[5]

But what about human rights? By the 1990s, arguments against FGM based on health risks and the negative medical consequences were being replaced with challenges to the practice on the basis of universal human rights (Shell-Duncan and Hernlund 2000:25). The discourse of universal human rights is itself an historical artifact that emerged at a particular time in history and that enshrines the normative values of certain European and European-descendant societies as rights for all people. The mobilization of rights-based arguments against FGM has resulted in a different sort of backlash, one in which opponents of the practice are characterized as neo-colonial and paternalistic (e.g., Morsy 1991). Importantly, under a rights-based approach it becomes impossible to justify an exemption for male genital modifications, no matter where they are performed.[14]

This is precisely the point that led me to my dissertation research on parental decision making about neonatal male circumcision in the United States. Female circumcision has received a disproportionate amount of scholarly attention relative to male circumcision. Scholarly research commonly focuses exclusively on female circumcision in societies where both are practiced (Bell 2005:127–128).[25] I believe that this situation reflects the relative normalization of male circumcision in European and European-descendant societies. When I say relative normalization, I mean that the practice is not questioned or prohibited even if it is primarily restricted to religious groups, as is the case in several European countries. Gender-based differential treatment is untenable if the issue is treated as one of human rights, and American cultural practices should not be exempt from critical and scientific inquiry.

Certainly, the case of male circumcision in the United States is not beset with the same kinds of political issues as the international campaign targeting female circumcision in African nations. The circumcision debate in the United States is an intra-cultural debate, free from the colonial baggage of the intercultural conflict that arises in various African contexts targeted by international organizations. In regards to this problem of cultural imperialism, what could be more appropriate than applying the human rights concept to the society responsible in large part for developing it? Despite their relationship in principle, the settings and contexts of childhood genital modification procedures in the United States and in African nations are so dramatically different that they are generally not comparable. I want to be clear that I see a whole series of fundamental differences between doing research in an African context and in a United States context, and I am not presuming to equate the two. But, if we take the practices of socially motivated genital modification out of context for a moment and look at the discourse arising around them, certain broad general characteristics appear that are common to both male and female circumcision, and it is these common characteristics that have led me to conduct my own research with a cultural relativist approach. Some of the broad trends that are apparent are a social expectation for genital modification, perpetuation of the practice by older generations who manifest it for their offspring, and the acceptability of the behavior within the community of practice. This last feature, of course, is the basis for a relativist approach and, to an extent, explains the backlash against interventionist responses to FGM mentioned above. The activist movement in the Anglophone world is not without its own backlash but, before I comment on that, I want to explain how I conduct my research into parental decision making on neonatal circumcision in the Unites States.

Making a scientific inquiry into human behavior is a challenge. Human behavior is incredibly complex, and trying to account for all of the various factors that go into any given individual's experience is daunting, to say the least. As an ethnographic social scientist, the methods I use for collecting this kind of information go well beyond simple surveys and questionnaires. For my doctoral research, I conduct in-depth interviews with parents about how they made the decision about circumcision, what kinds of information they relied on, their sources of that information, their interactions with various health care providers, and so on. Getting strangers to talk openly and honestly with me about these things can itself be a challenge, and doing this successfully requires the establishment of a rapport between me, the researcher, and they, the subjects.

In this case, "rapport" refers to the dynamic between an ethnographer and the informants. Building rapport is really about building relationships; in whatever context an ethnographer is working, they are always an outsider, and they are asking people to let them in. Before people will let you in, they have to trust you. Developing this kind of trust requires honesty, of course but, in my experience, it also relies heavily on a kind of intellectual humility. If I already know the answers, why am I bothering to ask people the questions? If I already know the truth about circumcision, why would I bother to ask them what they think? This brings us back to the notion of efficacy. My goal as an anthropologist is to understand people's behavior. If I want accurate information from people about their ideas, beliefs, and motivations for their behavior, then I have to make sure that they trust me enough and feel comfortable enough with me to share their true feelings and beliefs with me. Building good rapport is about developing good relationships with the people you are asking to divulge inner feelings. Challenging people's beliefs or claiming to know better than they do what is appropriate for their children are not effective strategies for establishing rapport. In my professional capacity as an anthropologist, it's quite simply not my job to tell people what they should or shouldn't do; but rather to try and understand why they do what they do.

Now, as I said before, I am personally opposed to neonatal circumcision. I find the rights-based arguments about age and choice to be particularly compelling. So, how do I reconcile this with my role as a relativist anthropologist? Well, whether or not we can personally see something from an objective standpoint doesn't always interfere with our ability to formulate objective questions. And I have sufficient faith in the notion of intellectual humility that I don't automatically take my own perspective for granted as necessarily accurate or correct, although it certainly is the position with which I am most comfortable. I see a fundamental and important difference between me having personal beliefs and me using my personal beliefs as a standard against which to evaluate other people's beliefs. This is what I call intellectual humility.

So, even though the debate over male circumcision is an intra-cultural debate unfolding within a particular society, to be scientific about our approach, we must step outside of the intra-cultural debate and treat people's beliefs and behaviors objectively, lest we fail to see past our own beliefs. This is a methodology; this is not a claim to real objectivity. I have my beliefs, but I am able to set them aside when it comes to investigating other people's beliefs. What would happen otherwise? Well, I mentioned a backlash against activism and anti-circumcision positions in the United States. The negative response generated by the international anti-FGM campaign is of course rooted deeply in the postcolonial context, from which the intra-cultural debate in the United States is free. The dynamic of intervention and backlash appears to play out nevertheless, and this is where I see the most value of a cultural relativist approach for my own research.

What about this backlash? Well, I recognize a distinct anti-anti-circumcision theme in several Internet resources on circumcision, and also in some of the scholarly literature. On the worldwide web, there are several sites that present themselves as direct responses to the anti-circumcision campaign; for example The Gilgal Society, which "acts as the sponsor of…an online resource set-up to provide correct information and to counter the lies, half-truths, and distortions with which anti-circumcision activists have flooded the web"[14]; Professor Brian Morris' website, circinfo.net, states that "unfortunately, the topic of circumcision has been made unnecessarily controversial because of emotive propaganda and opinions placed on the Internet by extremist anti-circumcision organizations" (Morris, 2006); the website circumcisioninfo.com explains that "these people have also saturated the Internet with their own, many times, [sic] fanatical points of view containing much misinformation regarding medical, sexual and psychological aspects of circumcision."[16] On the publicly written and edited Internet encyclopedia, wikipedia.org, an editor who is prominent on almost all pages related to circumcision states that "when I came to Wikipedia, I was horrified to see that anti-circumcision activists had authored pages on the subject of circumcision that read much like a crank website. The pages were incredibly biased and far from factual. I am working to resolve this problem, but it is far from easy."[17] Lastly, my personal favorite example of a backlash against the anti-circumcision movement, albeit one that may end up having a counter-productive effect: the Circumcision Independent Reference and Commentary Service, which can be found at circs.org, features a page that lists "notable circumcision opponents," a list that, in its entirety, includes: three pediatricians, a nurse, a historian, and Adolf Hitler.[18]

The point of these examples is to demonstrate that, regardless of the veracity of these sites' claims about the benefits and desirability of circumcision, there is a pronounced Internet reaction to activism. It is clear to me that this trend is not simply pro-circumcision, but anti-anti-circumcision. And there is at least one anthropologist who has been observing and reporting on the anti-circumcision movement in the United States, although he is clear that he does not find their arguments to be "morally compelling."[19]

What about on an interpersonal level? I personally alienated at least one potential informant by wearing an anti-circumcision T-shirt. His wife and I know someone in common, and she had agreed to participate in an interview on how they made the decision to circumcise their son, but after seeing me at a cookout wearing the shirt, he declined to participate, fearing criticism of his preference for circumcision. Another parent who volunteered to participate in my research had elected circumcision for their newborn and was referred to a pediatric urologist for an unrelated condition, whereupon she was met with outspoken criticism of her choice from the urologist, which she felt was highly inappropriate. In these cases, parents encountered a challenge to their beliefs, and reacted negatively. The tactics and strategies inherent in an interventionist approach to the issue create conditions in which people who feel that they are behaving appropriately when they choose circumcision will respond negatively to their decision being challenged. I was interviewing a mother who was opposed to circumcision, and she used the phrase "circumcision Nazis" to describe the zealousness of the anti-circumcision material that she had found on the Internet. So, even though this person agrees in principle with the message, she clearly does not like the way in which the message is being presented.

So, what are the implications for activism? That is for the activists to decide. My intention with this paper is to explain my approach as an anthropologist studying the practice of socially motivated genital modification. I am absolutely not claiming to know better than anyone else how to approach discussions with parents about their circumcision choices. I personally agree with a lot of what a lot of activists believe, but I can't seem to get to a place of intellectual righteousness with it, and I wholeheartedly agree with Rock Brynner, who said "as soon as we start judging someone, we stop understanding them."

References

1. Hodges FM 2001. The Ideal Prepuce in Ancient Greece and Rome: Male Genital Aesthetics and Their Relation to Lipodermos, Circumcision, Foreskin Restoration, and the Kynodesme. Bulletin of the History of Medicine, Fall;75(3):375–405.
2. Gowans, Chris 2004. Moral Relativism. *In* The Stanford Encyclopedia of Philosophy, Spring 2004, Edward N. Zalta ed. Electronic document, http://plato.stanford.edu/archives/spr2004/entries/moral-relativism, accessed August 14, 2006.
3. Allaboutphilosophy.org 2006. Moral Relativism: Neutral Thinking? Electronic document, http://www.moral-relativism.com, accessed August 14, 2006.

4. Moralrelativism.info 2006. Moral Relativism: Is Morality Relative to Individuals or Cultures? Electronic document, http://www.moralrelativism.info/index.html, accessed August 14, 2006.
5. Gruenbaum, Ellen 2001. The Female Circumcision Controversy: An Anthropological Perspective. Philadelphia: University of Pennsylvania Press.
6. Shell-Duncan, Bettina and Hernlund, Ylva, eds. 2000. Female "Circumcision" in Africa: Culture, Controversy, and Change. Boulder: Lynne Rienner Publishers.
7. Konner, Melvin 1990. Mutilated in the Name of Tradition. New York Times Book Review, April 15:5.
8. Gordon, Daniel 1991. Female Circumcision and Genital Operations in Egypt and Sudan: A dilemma for Medical Anthropology. Medical Anthropology Quarterly 5(1):3–14.
9. Mackie, Gerry 2003. Female Genital Cutting: A Harmless Practice? Medical Anthropology Quarterly 17(2):135–158.
10. Obermeyer, Carla Makhlouf 2003. The Health Consequences of Female Circumcision: Science, Advocacy, and Standards of Evidence. Medical Anthropology Quarterly 17(3): 394–412. 1999. Female Genital Surgeries: The Known, The Unknown, and the Unknowable. Medical Anthropology Quarterly 13(1):79–106.
11. Foster, George M. and Barbara Gallatin Anderson 1978. Medical Anthropology. New York: Alfred A. Knopf.
12. Semakula, John Kiwanuka 2002. Africa Health Analysis-selected Data from World Health Report 2002. Medilinks: Africa Health Report 2002. Electronic document, http://www.africa.upenn.edu/health/news.htm, accessed August 2, 2006.
13. Ahmadu, Fuambai 2000. Rites and Wrongs: An Insider/Outsider Reflects on Power and Excision. *In* Female "Circumcision" in Africa: Culture, Controversy, and Change. Bettina Shell-Duncan and Ylva Hernlund, eds. Pp. 283–312. Boulder: Lynne Reinner.
14. Androus, Zachary 2004. The United States, FGM, and Global Rights to Bodily Integrity. Paper presented at The United States and Global Human Rights. Rothermere American Institute at Oxford University, United Kingdom.
15. The Gilgal Society 2002. Circumcision — A Guide to a Decision. Electronic document, http://www.circinfo.com/guide_to_decision/index.html, accessed July 17, 2006.
16. Circumcision (Male and Female) Information Pages 2006. Circumcision Information Explained by Doctors, Medical Establishments and Researchers. Electronic document, http://www.circumcisioninfo.com/index_home_new.html, accessed July 17, 2006.
17. Waskett, Jake 2006. User:Jakew-Wikipedia, the Free Encyclopedia. Electronic document, http://en.wikipedia.org/wiki/User:Jakew, accessed July 17, 2006.
18. CIRCS.org (Circumcision Independent Reference and Commentary Service) 2005. The Circumcision Debate: Notable Circumcision Opponents. Electronic document, http://www.circs.org/reviews/debate/, accessed August 17, 2006.
19. Silverman, Eric K. 2004. Anthropology and Circumcision. Annual Review of Anthropology 33:419–445.
20. Morris, Brian J. 2006. Benefits of Circumcision: Medical Health and Sexual. Electronic document, http://www.circinfo.net/#why, accessed July 17, 2006.
21. Morris, Brian 2006. Benefits of Circumcision: Medical, Health and Sexual. Electronic document, http://www.circinfo.net, accessed August 15, 2006.
22. Morsey, Soheir A. 1991. Safeguarding Women's Bodies: The White Man's Burden Medicalized. Medical Anthropology Quarterly 5(1):19–23.
23. Boddy, Janice 1991. Body Politics: Continuing the Anticircumcision Crusdae. Medical Antrhopology Quarterly 5(1):15–17.
24. Sargent, Carolyn 1991. Confronting Patriarchy: The Potential for Advocacy in Medical Antrhopology. Medical Antrhopology Quarterly 5(1):24–25.
25. Bell, Kirsten 2005. Genital Cutting and Western Discourses on Sexuality. Medical Anthropology Quarterly 19(2):125–148

Medical Interventions on Women's Genitals: Historical Texts and Contemporary Discourse

Noel Gazzano

Abstract In Italian media and professional medical discourse, medical interventions on bodies are described in non-cultural terms and are presented as necessary responses to material problems considered objectively assessed. Far from being a mere linguistic process, this categorization is dense with meaning: it removes medicine from the realm of culture, transforming it from a sociocultural process to an historical object, a given. The controversy around a proposal for a ritual alternative to FGM in a Florence clinic triggered my reflection on the aforementioned medical view that organizes human reality into two conflicting shpheres: medical practices on one side and cultural actions on the other. In the context of a reflexive perspective on biomedicine, I analyze 19th-century Italian medical journals that consider female genital surgery as a cure for a variety of ailments; this analysis gives useful insight on the contemporary debate–alive in Italian biomedical environments–on the definition of medical practices in relation to interventions on patients' bodies and on doctors' roles in Italian society.

In January 2004, the mass media informed the Italian public that two gynecologists in Florence, Italy, had proposed to practice, in a local public clinic, a procedure to substitute infibulation, believing that local girls of Somali origin were at risk of undergoing FGM in Italy, or during a trip in Africa. The proposed operation consisted of a prick on the prepuce of the clitoris with a needle or a lancet, having previously covered the part with anesthetic cream (*Lidocaine*, *Prilocaine*). The proposal, often called "alternative ritual," quickly became a national case, spurring vivacious debates among the general public, but especially among immigrant women activist groups and medical doctors. It is on the latter that I will focus my attention.

In the medical system, the deontological notion of *proper medical action* was at stake, as well as the opportunity or not for the local healthcare system — which is public — of offering such a service. Could the so-called "alternative ritual" be performed in a medical environment?

During the years 2004 and 2005, I interviewed medical doctors, as well as members of the Italian bioethical committee and of the local doctors' organization;

Adjunct Professor of Anthropology, Instituto Lorenzo dè Medici and Marist College, Firenze, Italy

G. Denniston et al. (eds.), *Circumcision and Human Rights*,

moreover, I analyzed all the pertinent articles that appeared in local and national newspapers, and documents published by the Italian government on both male and female genital modifications. I noticed that, throughout the debate, the terms *ritual* and *cultural* were associated with practices that were defined as *inherently non-medical*, thus questioning the pertinence of the "alternative ritual" to the medical realm; in other words, certain practices were described through cultural categories, and considered in opposition with medical practices, which were implicitly characterized as having a non-cultural nature because they were considered *necessary* responses to *objective* and material problems.

In medical discourse: medical, necessary, curative actions on one side; rites, cultural, unnecessary, non-medical actions on the other.

Far from being a mere linguistic process, this categorization is dense with meaning: it removes biomedicine from the realm of culture and thus from the world of choices, possibilities, and relativity, transforming it in an a-historical object, a *given*. Biomedicine is described as being characterized by a specific *action* — observation — the objectivity of which is guaranteed by more and more precise technological devices, and by a specific *object* — the human body — considered observable or, better, knowable through observation.[1] It is precisely through this ideology of objectivity that biomedicine constructs and presents itself as a system of knowledge free of metaphors, a "copy of reality" without ambiguities or distortions, so that medical actions are *objectively* alien to the realms of culture and society. Citing Hernáez, Ángel Martínez: "Rendere Visible l'Invisible. L'antropologia e la Trasparenza del Potere Biomedico" (2000, 9–10:9–34).

Indeed, many Italian doctors recognize that, as individuals, both they and their patients "have their own culture" and, to a certain extent, they also recognize the "culturality" implicit in medical practice, acknowledging the efficacy of alternative medicines; this recognition of variability transforms biomedicine in a relative matter: in this perspective, biomedicine becomes an alternative to other medical traditions. Nonetheless, what isn't put into question is the sociocultural nature of medical diagnosis and cure *per se* (Pizza 2005:127)[2]; when this happens, the "culturality" is identified with a lack of knowledge that time remedies; this critique is referred mainly to the past, and the faith in constant scientific progress based on growing knowledge through in-depth observation undermines the possibility of a reflexive perspective.

In this context, (contemporary) proper medical actions are defined as *necessary* and *curative* interventions on bodies, and it is against this specific definition that the Proposal was evaluated.

One of the main issues in the debate was precisely the (alleged) conflict between the nature of the Proposal — "a ritual" — and the environment in which it was proposed — a hospital; supposing the "alternative ritual" was accepted as part of the services offered by the Italian healthcare system, parties involved would have had to reconcile an orthodox definition of medical interventions on bodies with the practice of such procedure. In the words of one of my informants, a member of the Italian bioethical committee: "We [doctors] cannot put on the sorcerer's hat!"

Some noticed that the Proposal can't be considered a cure because it doesn't affect an ill part of the body[3]; nonetheless, the proponents argued that it could be considered a form of *preventive* medicine, seeing as how it avoids further damage.

On the contrary, an informant of mine sharply noticed that this definition doesn't apply, given the absence of a pathogen agent in the strict sense of the word, while in this case the agent is simply human will; the intervention should modify that will, and not a third, unconnected, element — the child's genitals — (and, apparently, legal action isn't sufficient to modify that will).[4]

Interestingly, the medical practice of what Italian governmental documents call male "ritual circumcision" wasn't considered a precedent[5]; the use of religious argumentations, and the controversial supposition of health benefits of male (Jewish) circumcision, can be read as an attempt to demonstrate *a posteriori* the coherence of this practice with the nature of medical action, given that doctors do perform it.

Indeed, male neonatal circumcision blurs the definition of canonical medical interventions on bodies, highlighting how the realms "cultural," "traditional," and "medical" often overlap.

Male circumcision is a "gray area" of medical practice, in which the words "social" and "cultural" emerge strongly, and it is precisely analyzing such gray areas that we can go beyond biomedicine's seeming uniformity and clarity of boundaries, highlighting its *processual* and *productive* nature.

I will concentrate on one such gray area of medical practice: female genital modifications.

Through an anthropological analysis of biomedical discourses on FGM, on the Proposal, and on nineteenth-century genital modifications in Europe, I suggest that the definition of proper medical action is an ongoing process that can be explained, correlating it to the specific socio-cultural context in which it occurs[2,6]; this ongoing process is concealed by the use of a language that communicates neutrality, necessity, and objectivity. Moreover, I state that this ongoing definition of proper medical action is a fundamental aspect of the constant definition and redefinition of biomedical identity through its boundaries, which characterizes biomedicine at once as a sociocultural process and product; conversely, if medical practices are always a social matter, at the same time, social matters often become medical practices: this process is called *medicalization* (Conrad 1992), and in this paper I interpret the aforementioned cases of medical female genital modification precisely as cases of medicalization.

Throughout the nineteenth century, the female internal and external genitals were object of close examination by specialists who dealt with a variety of ailments, which involved aspects of human life that we could call physical, behavioral, and emotional; these ailments were grossly defined "hysteria," but the existence of such a nosological category was the object of great debate in the scientific journals of the time.

I analyzed three journals: the *Archivio Italiano per le Malattie Nervose e più particolarmente per le Alienazioni Mentali* (the *Italian Archive for Nervous Maladies and Especially Mental Alienations*), from its first issue in 1864 to the last one in 1891, when it became part of the *Rivista Sperimentale di Freniatria e Medicina Legale* (*Experimental Journal of Phreniatry and Legal Medicine*), which I researched from 1875, its first year, up to 1886. Finally, I studied the *Annales Médico Psychologiques*, from the first issue in 1843, up to 1861; this journal was linked to the Salpêtrière, the hospital made famous at the end of the century by Charcot's treatment of hysteria with hypnosis.

For reasons that will soon be clear, I coded for articles that included the words: hysteria, epilepsy, convulsions, nymphomania, erotomania, genitals, vagina, clitoris, uterus, ovaries, sexuality, women's maladies.

The emergence of hysteria and its manifestations were linked to convulsions, epilepsy, anemia, nymphomania, erotomania; ovaric, uterine, and menstrual disorders, following the ancient Hippocratic theory; or nervous disorders, variously linked to the uterus — when the ancient theories were updated with the latest pathological anatomy discoveries (*Annales* 1847, 9:105–112; *Annales Medico Psychologiques*, Paris); diet; inheritance; contagion (*Archivio Italiano*, XXIV, 1887, 263:288; *Archivio Italiano per le Malattie Nervose e più Particolarmente per le Alienazioni Mentali. lOrgano della società freniatrica Italiana*. Ed. by Andrea Verga and Serafina Biffi. Milano, Chiusi, Rechiedei); emotional distress, shock, intense feelings, sudden fear; a variety of behaviors, including sexual ones; and, not last, a specific "sanguineous temperament." The illness could manifest with pale complexion; convulsions; genital, pelvic or back pain; agitation, fear, preoccupation, and strange fantasies. It could be identified during an autopsy, assessing the size and color of ovaries, as well as the state of the uterus and the color and shape of certain nerves. The cures were as varied as the descriptions: a rich diet — and at times a poor one; blood-letting, leeches; cold or warm baths, at time with the addition of different substances; a not well defined "moral cure," which in those days simply meant "not physical"; a variety of pharmaceuticals; and, finally, a direct intervention on the genitals.

This last form of cure ranged from touching the ovaric region in various ways, to actually removing the ovaries; as mentioned in an 1888 article from the *Archivio Italiano* (XXV:115), so-called "castration" was believed to be a cure for "hystero-epilepsy, epileptic phrenosis and hysterical ailments of menstrual origin." In the articles I examined, clitoridectomy also appeared: in 1847, the French *Annales* state the utility of this operation to "cure" masturbation, which was called *maladie*, illness (10:464–465).

A specification: research on hysteria, nymphomania, and other gender-based illnesses cite widely the clitoridectomies performed in the second half of the nineteenth century by British doctor Isaac Baker Brown, who believed that many obscure nervous diseases such as hysteria, epilepsy and catalepsy, as well as the masturbation which gave rise to them, could be cured by this operation. What studies don't cite is precisely the fact that female genital operations were present even in France and Italy, and, as we can see reading the 1847 article I just cited, clitoridectomies were performed in France before Baker's publication.

To confront historically the medical treatment of hysteria means to enter a slippery world: physiology, temperament, behavior, and emotions merge in descriptions that leave the researcher with the difficulty of grasping their conflicting meanings. Nonetheless, analyzing, diachronically, these articles, it is possible to "make sense" of the medical discourse on hysterical ailments, gaining, as mentioned above, important insight on the definition of medical action, and on medicalization, that is the role of biomedical thought and practice within social issues.

In the descriptions of hysteria as a nosological entity, and in the anamneses, diagnoses, cures, and outcomes of patients, we can identify a *fil rouge* of elements that appeared roughly outlined in the 1840s, and were later described as prevalent features of these ailments. We can retrace the development of such elements, following chronologically the Italian and French publications: while in the 1840s hysteria was described mainly as a convulsive illness, with time it became more and more a matter of temperament and (often sexual) behavior, the latter defined in terms of morals and social acceptability.

This development of the notion of hysteria runs parallel to its characterization as a female ailment: if in the 1840s and 1850s hysteria is mainly a convulsive illness, and as such it can be experienced by both men and women, with time, it will be described as expressed by a peculiar temperament and by character traits considered to be *feminine* and linked to the female body (the presence of uterus, ovaries, and menstruation).

If hysteria at first was identified with convulsions — that could be easily diagnosed, although at times they were mistaken with epilepsy — it gradually became an illness characterized by multiple pathological forms, in which the boundary between body and behavior, nerves and education, psyche, and character, became more and more tenuous. This opened the possibility of linking illness and socially deviant behavior, the definition of which could change, given the elusive way in which the illness itself was constructed.

In the 1840s and 1850s, women diagnosed with hysteria and convulsions were *also* said to be quite emotional, impressionable, and with a "sanguineous temperament" (and at times the "exaggerated sensibility" was described as accompanied by a lack of intelligence); these peripheral elements become subsequently prevalent, so that hysterical patients were described as women with a certain character and lifestyle — and, with a gradual shift, precisely that character and lifestyle became at once cause and symptom of illness.

I therefore suggest that hysteria became, throughout the second half of the nineteenth century, a "place" for the construction, definition, and maintenance of what was considered proper female behavior, through the identification of social deviance as illness; in brief: hysteria became the place for the construction and maintenance of a specific gender identity.

The development of a specific female identity can be traced throughout these texts, in particular where illness was described and diagnosed in reference to an alleged "normality"; if, in an 1847 case of a young hysterical woman (*Annales* 1847, 9:287–290), normality was defined by size of ovaries, age of first menstruation, and color of left intercostal nerve, soon after other elements entered the scene, and in particular the degree of emotional, physical, and especially genital sensibility. And, it is interesting to notice, the same article, making reference to a former case cited in the *Annales*, stated that sexual activity enhances illness.

The "hyper-excitability" of epidermis, ovaries, and genitals is accompanied by the notion of an "exaggerated sensibility" of these women: the notion of "exaggeration" appears, and will have great part in defining the limit between normal and abnormal, healthy and pathological. If, in the first half of the nineteenth century, the exaggerated sexual drive of nymphomaniacs and erotomaniacs, as well as onanism, were distinct from hysteria, that was identified with convulsions and nervous disorders and was cured with drugs, during the second half of the century, these domains somehow overlapped. It is possible to identify an interesting passage in an 1847 article published in the *Annales* (9:110), which states that hysterical ailments of the central nervous system can be cured by an *effort of will*: we thus enter the social realm of behavior, with the suggestion that, if the rational control of the self can be a cure, its opposite, exaggerated emotion, is negative — and indeed it will soon become *maladie, maladie féminin.*

Although it is quite dubious that it represents a common trend in British medicine, in 1867, the *Medical Times and Gazette* published an article that criticized Brown's procedures, and that made statements that conflict with what we find in many articles published in contemporary Italian and French journals: "It is an ethical offence, in the first place, if the Practitioner who is consulted for any common complaint, say hysteria or fissure of the rectum, set himself to consider whether or not the patient is guilty of immoral practices, which have nothing to do with the case before him."[7] As we saw, the anamneses and diagnoses in these journals growingly focus on personality traits and behavior.

Moreover: "Affirming then, in the first place, that the very entry of thoughts of pollution into the Practitioner's mind respecting his patients is an offence of the deepest dye, this offence is aggravated by the kind of evidence which the clitoridectomist is taught to accept as proof of his patient's guilt. That evidence consists, partly, in certain physical signs detailed in Mr. Baker Brown's book — a 'peculiar straight and coarse hirsute growth,' a peculiar follicular secretion, and other phenomena detected by inspection, which are as frivolous as they are disgusting."[7] And, indeed, in the second half of the nineteenth century, anamneses included precisely this: description of detailed physical features, which include hair on the body, size and shape of genitals.

It is interesting to notice how the physical and behavioral features on which doctors concentrated their attention were present in anthropological texts of the time (Lombroso: one for all); indeed, the aforementioned *Journal of Experimental Phreniatry* said in the heading "in relation to anthropology and juridical and social sciences."

Beginning in the nineteenth century, anthropology, sociology, demography, urbanism, and hybrid fields, such as social hygiene and social medicine, identified the social domain as their object. Their goals were to define its identity, and diagnose, cure, and prevent the diseases that threatened it. The social body added another reified entity to Western discourses on itself: "It would be a mistake to regard the social as a natural category, as something given, as a scientific object waiting to be discovered in the nineteenth century."

Nonetheless, "the cultural and historical peculiarity of the ongoing attempts to know social bodies has, until recently, escaped the attention even of anthropologists, for whom the population, society, and 'the social' have too often appeared as transcultural, trans-historical objects" (Horn 1994:5). The search for the "true" human (modern) body meant, therefore, also the diffusion of uniformity. And it is interesting to notice that it was precisely in those years — the 1890s — that modern aesthetic surgery was born.

This anthropoietic process took place, intervening especially in the sexual realm. Women's sexuality is the product of the sedimentation of notions that were developed, among others, by anthropologists and doctors. Biomedicine showed all its pervasiveness and its active productivity: far from being merely a practical solution to physical ailments, it was a place for giving form to gender identities.

Anthropology defined racial norm, and plastic surgery molded racially fit bodies; criminal anthropology and psychiatry defined deviant women.

Medical science identified gender roles, diagnosed deviance, and embodied gender stereotypes — literally inscribing those ideals in women's flesh.

References

1. Good, Byron J., 1999: Narrare la Malattia. Lo Sguardo Antropologico sul Rapporto Medico-Paziente. Torino: Edizioni di Comunità.
2. Pizza, Giovanni 2005: Antropologia Medica. Saperi, Pratiche e Politiche del Corpo. Roma: Carocci.
3. Abi, Laila, 2005: "Nel Nome della Tradizione, Ti Purifico," in Busoni, Mila and Elena Laurenzi 2005: *Il Corpo dei Simboli. Nodi Teorici e Politici di un Dibattito sulle Mutilazioni Genitali Femminili.* Firenze: SEID. and Laurenzi 2005:161–172.
4. Stephens-Davidowitz, Lauren, 2003: "Traditions Die Hard. Can Arrests Stop Female Genital Mutilation?" in Ghanaian Chronicle (Accra), 19 November, 2003.
5. Comitato Nazionale per la Bioetica, 1998: *La Circoncisione: Profili Bioetici.* Presidenza del Consiglio dei Ministri. Dipartimento per l'Informazione e l'Editoria.
6. Lock, Margaret and Deborah R. Gordon (eds.) 1988: Biomedicine Examined. Dordrecht, Bston and London: Kluwer Academic Publishers.
7. Anonymous, 1867: "Clitoridectomy and Medical Ethics," in Medical Times and Gazette (London) 1867(1):391–392.
8. British Medical Journal, 1866: "Review of 'On the Curability of Certain forms of Insanity, Epilepsy, Catalepsy and Hysteria in Females, by Isaac Baker Brown,'" 28 April:438–440.
9. British Medical Journal, 1867: "Report of Obstetrical Society Meeting to Investigate Isaac Baker Brown's Practice of Clitoridectomy," 6 April:395-410.
10. Brown, Isaak Baker, 1866: On the Curability of Certain Forms of Insanity, Epilepsy, Catalepsy and Hysteria in Females. London: Robert Hardwicke.
11. Busoni, Mila and Elena Laurenzi, 2005: Il Corpo dei Simboli. Nodi Teorici e Politici di un Dibattito sulle Mutilazioni Genitali Femminili. Firenze: SEID.
12. Comitato Nazionale per la Bioetica, 1998: *Problemi Bioetici in una Società Multietnica.* Presidenza del Consiglio dei Ministri. Dipartimento per l'Informazione e l'Editoria.
13. Comitato Nazionale per la Bioetica, 2001: Bioetica Interculturale. Bozza del Parere del CNB per la discussione. Roma, Luglio 2001.
14. Hernáez, Ángel Martínez, 2000: "Rendere Visibile l'Invisible. L'antropologia e la Trasparenza del Potere Biomedico," in *AM, Rivista della Società Italiana di Antropologia Medica*, 9–10:9–34.
15. Lock, Margaret and Deborah R. Gordon (eds.), 1988: Biomedicine Examined. Dordrecht, Boston and London: Kluwer.
16. Conrad, Peter 1992: "Medicalization and Social Control, in Annual Review of Sociology, 18:209–32.
17. Horn, David G. 1994: Social Bodies. Science, Reproduction and Italian Modernity. Princeton: Princeton University Press.

1996–2005 — Ten Years of Merka's Alternative Ritual in Somalia

From "Sunna Gudnìn" to "Gudnìn Usùb" ("The New Rite")

Mana Sultan Abdurahman Ali Issah[1] and Pia Grassivaro Gallo[2]

Abstract The following describes Merka's temporary compromise that provides for the gradual passage towards the total elimination of female circumcision in Somalia.

Said passage incorporates the following features: abolishing any invasive intervention on the female genitalia; maintaining the presence of a medical practitioner, such as a nurse or traditional obstetrician, to give the impression that the clitoral puncture has taken place; conserving the "rite of passage" and associated celebration.

Between 1996 and 2005, this compromise was put into action in the phases articulated in the following table:

Years	Alternative rite	Female children involved
1996–2000	*Sunna Gudnìn*	1,300
2000–2002	*Gudnìn Usùb*	1,080
2003–2004	Presence of medical practitioners (no intervention)	562

Merka's alternative ritual is now employed in 32 villages of the Lower Scebelli, between this river and the coast of the Indian Ocean. It has been successfully applied in nearly 3,000 cases.

Excissory Traditions in East African Countries

The Djibouti Territory

The African strata of the population of Djibouti, which is also composed of Arab, Indian, and European groups, refer to the *Afar* and the *Issa*.

[1]WFL-Merka (NGO), Somalia

[2]Padua's Group on FGM, University of Padua, Italy

G. Denniston et al. (eds.), *Circumcision and Human Rights*,

The former are of Ethiopian origin (*Danàchili*), with the others are included a Somali pastoral group known as the *Dir*.[1] Both groups are Muslim, and traditionally practice female circumcision.

According to a 2002 study commissioned by the Djibouti Ministry of Health, 93% of women between the ages of 14 and 45 giving birth in cities had been subjected to a severe form of female genital mutilation (Type 2 or 3).[2, 3] A more recent study confirms the finding,[4] estimating the value closer to 98%.

The population recognizes two forms of female circumcision called *Pharaònigha* and *Sunna*, distinguished by differing degrees of severity in their side effects. Today, while the former corresponds with WHO type 3,[2] translated into western vocabulary as infibulation, *Sunna* does not match type 1 as would be expected from its nomenclature,[2] but has come to encompass all interventions that leave the vagina open. In practice, it is difficult to tell which type of intervention a young girl has undergone, as an unintentional closure of the vagina may result from having her legs tied together.

From a religious perspective, the two interventions are clearly differentiated: *sunna* is seen as a compromise between the many and contradictory references made in the *hadiths* of the prophet Muhammad on the subject of culture and tradition in Djibouti, while the *Pharaònigha* is considered to be part of a pagan legacy inherited from the ancient Egyptians, and is therefore removed from Islamic law, even though it is never explicitly or publicly condemned.[3]

The religious culture of the Imam does not normally allow any sustained debate on FGM, and most accept the standard interpretation declaring that the ritual is "recommended or tolerated." One must also remember the orthographic confusion that can arise between "*sunna*" (the attenuated form of female circumcision) and "*Sunnah*" (a body of Islamic traditions that make reference to the Prophet and his followers).

The same ambivalent behavior emerged among religious leaders during the 2001 Congress of Argesia.[5]

It was with such equivocation that the recent Sub-Regional Congress of Djibouti Against FGM took place (February 2nd–3rd, 2005), organized by the government of Djibouti and the international organization "No Peace Without Justice," the objective of which was to obtain a religious and political consensus against MGF.

After two days of tense discussion, the religious attempt to declare "partial excision of the clitoris legitimate on the condition that it be done by a specialist" was made so vain by the proud female protest that the Minister of Religious Affairs of Djibouti was forced to declare that "in the name of kind and clement God, that phrase must be erased from the final document." Therefore, Djibouti became the ninth country, and the first of the Horn of Africa, to ratify the Maputo Protocol on the rights of the African woman, initiated in 2003 in Mozambique (AA.VV. 8 February 2005).[6] It now remains to be seen to what degree the population, only a small part of which is urban, and much of which consists of nomadic nuclei that live far from any connection to the media, will manage to change their actual behavior as a result of the social intervention.

Rural Sudan

To delineate the recent evolution of the excissory tradition in Sudan, we refer to the consolidated experience of E. Gruenbaum as exposed in a recent article reporting a field study carried out in seven rural Sudanese communities.[13] Through interviews and discussions, he verified the continued presence of extreme female resistance to abandoning the tradition of female circumcision, especially regarding the vaginal closure; a modification that alters many daily functions, but gives the bearer a sense of adequacy.

"One cannot leave a woman open, like the road that leads to Ondurman...you need to cover (the opening) a little bit...a girl cannot be dirty, open, and having an intimate odor...the stream of urine cannot make noise or create a little fountain on the sand where it falls."

The compromise reached to avoid such "abnormal" behavior is to "remove the clitoris and close her a little (or often more than a little) above the urethra so that there is a *sunna*." This term, therefore, becomes estranged from the WHO definition (1996),[2] making it almost indistinguishable from infibulation, which has been in turn attenuated with respect to its history. The boundary between the two is thus becoming less certain, as the two interventions reciprocally approach each other, creating a modification in a "no-man's land" that is given the name *Sunna*.

Completing the picture is the fact that the use of a "sacred" vocabulary obtains the result that the intervention loses its prohibited, sinful aspect, slipping instead under the umbrella of the traditions related to the *Hadiths* of the prophet, although not accepted by all as genuine.

Therefore, *Sunna* circumcision in rural Sudan has taken on a positive connotation, accepted by religion, that does not interrupt the daily behavior of "good women."

Ethiopia

Female circumcision, which has a prevalence of around 75%, concerns all the populations of Ethiopia's primary regions (Ethiopia: *Regional States*, page IV). In particular, there are five groups (*Somali, Afar, Harari, Oromo*, and to a certain degree the *Bertu Jebeluwi*) that are infibulators, while the populations of the south, the *Tigrini* and the *Amhara*, practice the attenuated forms, types 1 and 2.[2]

Local linguistics are quite poor with regards to the different typologies of interventions, as many do not have specific terminologies.

In the last several years, some ethnic groups (the *Berta* and *Harari*) have reduced their interventions from infibulation to type 1 (*sunna*); a limited but similar trend is emerging for the *Afar*, *Somali*, and *Oromo* as well. Additionally, there have been several cases reported of the ceremony without any intervention actually taking place among the *Harari*.

The evolution towards attenuated forms, emerging from focus group discussions, is promoted by popular decisions supported by political and religious leaders. There are not, however, in-depth surveys that fully evaluate such initiatives. In stark contrast to this, there is no shortage of people who strongly oppose this idea, and campaign against the abolition of the practice.

(All of the information reported above was gathered from AA.VV., *Old Beyond Imaginings. Ethiopia's Harmful Traditional Practices*, 2003).[7]

The Population of the Lower Scebelli (Somalia)

The populations of the Lower Scebelli are mixed: there are three main ethnic components. The *Bimal* nuclei, which refer to the *Dir* ethnicity; the *Sab* people, who are the lightest-skinned, and the *Dighil* of the Scebelli,[8] locally called the *Rer Hamar*, the negroid *bantù* groups.

The last two groups in particular are characterized by a mixed economy of "agro-pastoral character that has evolved through specialization in irrigation agriculture and animal husbandry that prefers cows to camels, both of which testify a change towards substantial systems."[8] Along the river there are agricultural villages of the *bantù*.[9] In the wooded areas across the river there are home bases of pastoral nomad groups.

The survey on female circumcision in Somalia carried out in the 1980s[10] found that this region had the highest incidence of non-infibulation with respect to the rest of the country, at slightly over 50%. From the data gathered in the field, it is clear that the Negroid population nuclei that live along the river mostly practice infibulation, a tradition that has surely been influenced and reinforced by frequent contact with the pastoral populations on the other side of the Scebelli.

The populations dispersed along the coast and in the urban areas of Merka have a more open mentality, resulting from frequent dealings with Arab maritime traffic, and use a different excissory tradition: the "true sunna," recognizable as Type 1 in the WHO classification system of female genital mutilation.[2]

Merka's Alternative Rite

From 1996 to 2000

In 1996, in the Merka district of the Lower Scebelli, a celebration and ritual alternative to infibulation was conceptualized, proposed, and organized by the first author, who had been involved for many years in the Non-Governmental Organization "Water for Life." The rite, called *Sunna Gudnìn* (described various times: Grassivaro Gallo, et al., 2001,[11] 2004[5]; Grassivaro Gallo, 2002[12]) proposes an attenuated intervention on the clitoris (WHO Type 1),[2] carried out within the cultural context of the traditional infibulation. The rite maintains the following details: the tent where the rite takes place; the clothing of the young girl, who is completely shaved;

the affectionate presence of family and friends; the celebration, complete with dancing and singing by the women; the moment after the intervention during which the young girl is seated on the ground with her legs tied together in parallel.

Two teams carry out the ritual. The religious component, led by a *sheik*, acts first, and is followed by a medical team, which carries out the actual intervention touching only the prepuce of the clitoris. No sewing occurs.

The rite involves three phases: *persuasion* of the mothers of at risk girls, who will later publicly accept that the alternative rite be performed on their daughters; *the celebration*, including the scarification intervention on the clitoris (*sunna*) as carried out by a medical team using antibiotic and anesthetic injections; and *follow-up* on the acceptance of the new ritual, carried out by medical teams in the homes of individual girls by interviewing their grandmothers, who are the custodians of the tradition.

The first ceremony structured as described was held on 3 November 1996 in a village called Ayuub and included young girls from nearby villages as well, after notifying the teachers at schools where the WFL organization was present. The *sunna gudnìn*, carried out regularly a few times a year, diffused into about 30 villages. In the first 4 years, 1,300 young girls, who represent nearly a third of those who attend school in the Merka district, have accepted it.

From 2000 to 2002

It was decided that an even more attenuated intervention be proposed to the mothers. The New Circumcision, given the name of *Gudnìn Usùb*, expanded to another 1080 young girls. In the "test" villages of Ayuub and the Internal Displaced People zone (which collects refugees from the Upper and Lower Giuba River area), situated in Merka's periphery, the intervention was gradually reduced to a symbolic act carried out with a medical stick (or a sterilized injection needle in cases where one was not available), which just scrapes the clitoris.

The program is carried out in six centers, where all of the young girls in the district that have accepted the new right assemble.

The *Gudnìn Usùb* project is articulated in four consecutive phases: awareness building, medical intervention, celebration, and feedback.

Awareness building is carried out by the religious team, which takes action a week before the ceremony by organizing a *shir* (assembly) in the villages in which the main figure is still the *sheik*. Mothers, grandmothers, and teachers attend these assemblies. School children are asked to draw the major risks associated with infibulation (*Gudnìn Fircooniga*), and frame them with significant sentences. Figures 1–5 are examples of the drawings produced, and were framed by the following comments:

> Urine cannot come out; it's the Pharaonic Circumcision's fault. Pharaonic Circumcision causes serious problems when giving birth. A girl who has undergone Pharaonic Circumcision is never happy when she gets married. Mommy, don't cut a piece of my body. Pharaonic Circumcision causes problems during the first period.

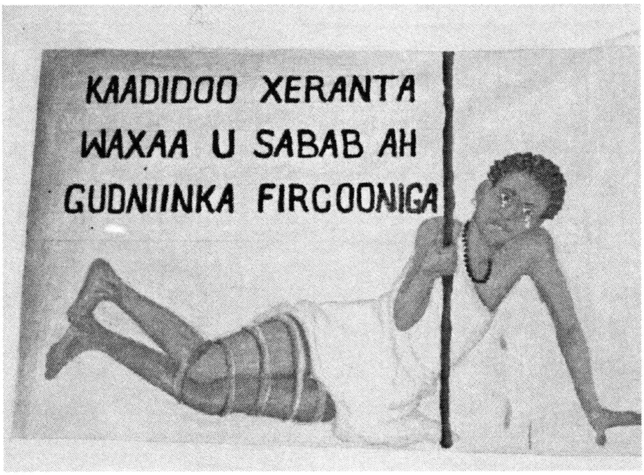

Fig. 1 Urine cannot come out; it's the Pharaonic Circumcision's fault

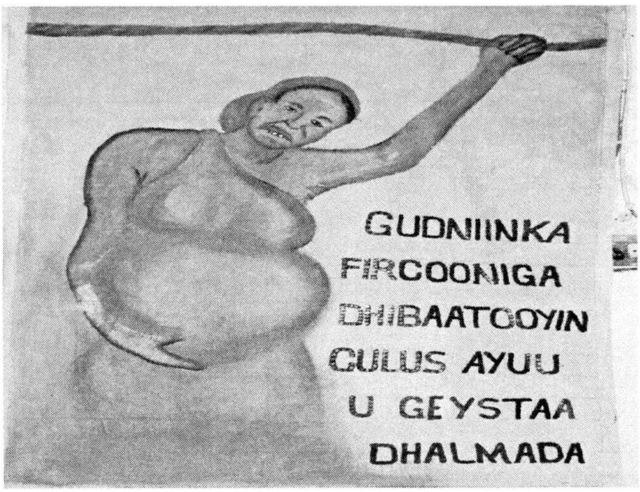

Fig. 2 Pharaonic Circumcision causes serious problems when giving birth

These drawings are shown at the aforementioned assemblies.

The medical intervention consists of a puncture of the clitoris by a medical stick or sterilized needle, which causes the loss of only a few drops of blood, and allows for the anesthetic and anti-tetanus measures to be eliminated. A medical team composed of 1 doctor, 2 nurses, and 12 community health workers (4 men and 8 women) carries out this phase. All team members have attended a specialized course organized by the WFL ONG.

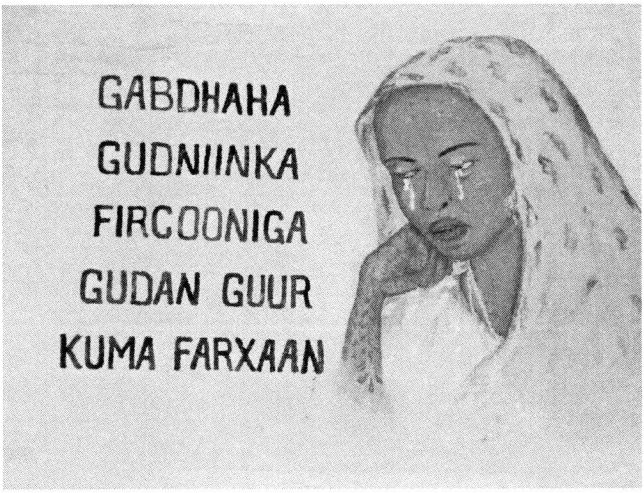

Fig. 3 A girl who has undergone Pharaonic Circumcision is never happy when she gets married

Fig. 4 Mommy, don't cut a piece of my body

After the intervention, the young girl stays at home with her legs tied together in parallel for around ten days. The mothers are given pills to provide every day of the "convalescence" period, which are claimed to "ease the pain." Truthfully, these pills are vitamins, not analgesics, as the girls have not been infibulated, and thus require no such medication. Such medical behavior, however, reconnects the alternative rite with traditional infibulation, reinforcing a sense in the subject and community that the tradition has been fulfilled.

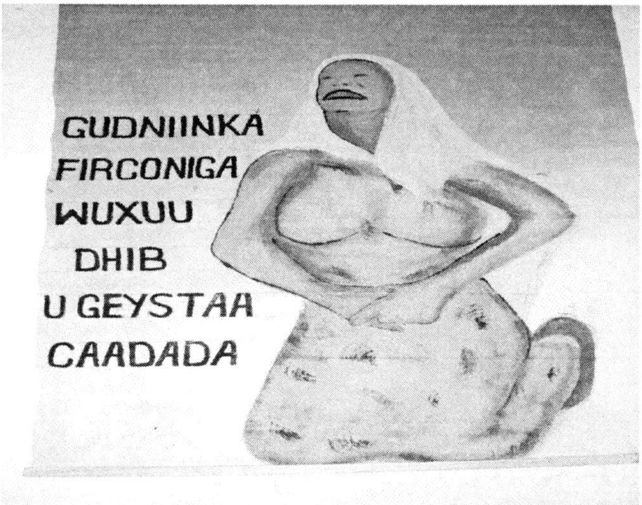

Fig. 5 Pharaonic Circumcision causes problems during the first period

The celebration of the new rite is conserved as an important cultural element, during which the young girl is given new clothes, sweets, and drinks. At the end of the celebration, the mother presents her daughter as initiated into the community, and declares her choice of *Gudnìn Usùb*, or circumcision according to the new rite. The girls stay home from school afterwards to underline that the intervention has occurred.

The feed back phase is carried out by social workers who conduct surveys in the families and communities where the rite has been accepted, to be sure that the alternative practice has not created marginalization or social discomfort. Women are invited to promote the new practice within their social circles.

During 2002, lack of funding caused WFL to carry out *Gudnìn Usùb* rites only once a year, when the schools close. The scholastic initiative and many schoolgirls were thus not involved. Every village organized itself autonomously, asking WFL only for a medical team and a contribution for the celebration costs.

From 2003 to 2004

An attempt to eliminate intervention on the clitoris was made. The experience involved 562 female children left completely intact in the villages of Ayuub, the surrounding area in the Merka district, and the refugee area.

Since 2002, Mana began another intervention in the city of Ayuub, where she resides, convinced that the moment had arrived to try for a complete eradication of the practice.

After many meetings and continuous discussions with the population with which she is closest, she succeeded in convincing the group of young mothers to not perform any intervention on their daughters.

Mana proceeded along the following steps. A secret agreement is made with the mother of a newborn, in accordance with which an obstetrician comes to the family's home to simulate an intervention. She will not actually perform any intervention, as agreed upon with Mana and the mother, but the community will see the child as circumcised at birth. As the girls grow up, they are told that the intervention was performed during the first days of their lives. Only when they are old enough to understand will they learn the whole truth: "I told my daughter that she had been circumcised when she was still very small. Only in time did I tell her the truth: she had not been touched" (Mana). In these cases, no ceremony takes place.

Support Activities

Some ex-orphans taken in by WFL who are now teachers (42 total) provide *education of mothers and female children* in the schools of different villages, recounting their own experience as non-infibulated girls. There young women are vibrant examples that demonstrate, especially for the elderly who are more resistant against changes in tradition, how wrong it is to believe that not infibulating a girl will generate problems for her future marriage.

Afternoon sewing and crafts classes are also organized in the villages, which are often used as occasions to persuade mothers against infibulating their daughters.

Meetings with the participation of infibulated women are also held, where they discuss the consequences and problems that result from the intervention.

According to Mana, "*It is now possible to talk about these subjects, which were taboo before. We have a saying, 'When you change something traditional, you attract the wrath of god.' At the beginning, I could not say 'Let's change the tradition,' because the mothers would have answered that it would have caused a curse from god, so I had the sheik say it instead.*"

Comments and Conclusions

The bibliographical analysis presented as an introduction elucidates that, in East African countries, there has been a loss of linguistic and semantic connotation of the word *sunna*, which originally had a more restricted definition as an attenuated intervention.[2] The term has been used to indiscriminately indicate every female genital intervention, but no longer means anything. This is quite probably due to an attempt by different populations to apply a sacred reference to a practice that is seen as noxious, but continues to be socially approved of.[13] This situation coerced Mana into changing the name of Merka's alternative rite from the originally proposed *sunna gudnìn* to *gudnìn usùb*, or "new rite."

Specifically, the alternative rite proposed should be viewed as a transition phase towards the complete eradication of excissory customs, accepting it temporarily as the lesser of two evils.

Two focal points of the passage are compromise, and gradual change.

The *compromise* is temporarily accepting an attenuated intervention that passes through two levels of attenuation: from the scarification of the clitoris for *sunna gudnìn* to a simple puncture of the clitoris for *gudnìn usùb* before taking the final step to eradication (or at least current suspension of all interventions), now maintaining the simple presence of medical personnel to protect the female child, while she is still young, with the premise of the rite being performed although it never was.

The *gradual* nature of the system underlines how attenuating the tradition in phases can result in ultimate abolition of the practice.

It should be noted that Mana, from the proposal to the procedure of the alternative rite, serves as a charismatic figure who belongs to the culture concerned, and is thus able to involve a whole village in a *shir*, the protagonists of which are the women; men; socially important figures like religious leaders, teachers, and obstetricians; and even school children, through their artistic representations. No western element of any title could hope to have a similar impact.

The rite conserves the frame of the traditional infibulation intervention to the highest degree possible, and includes the celebration at a salient and gratifying moment for the young girl.

From a social point of view, one sees the importance of a mother's assuming responsibility for choosing the new rite when she presents her initiated daughter to the community. On this point, Mana retains that "*if the group that did not undergo any intervention succeeds in not being eliminated by other women, it will later be possible to reveal the truth when they are old enough to understand.*"

A group of subjects that testify having broken with the tradition is considered by Mackie (2000)[14] to be a fundamental step towards successful eradication of the practice. The author believes that it is indispensable for the "positive deviants" to form a critical group of individuals in the community that publicly denounces the custom. Mana understood this risk by instinct and experience without any need of bibliographical reference.

To close, there is one last consideration. In the western world, we are used to considering the negative aspects alone of infibulation, but the ritual within its cultural context (the high value attributed to the young girl, the care and attentiveness that the adults show her as a new initiate in society, etc.) give a young woman the possibility to learn to behave herself appropriately in all of life's situations. Now, Mana's alternative rite conserves this cultural patrimony, which is channeled into the celebration, the protection of the girl from marginalization by her peers during school, and the future projection of professionalism acquired through study, which allow "open" young women the opportunity to marry.

The first girls from Ayuub to have *sunna gudnìn* were not socially marginalized, and through the years they have formed their own families and become mothers. There social and professional profile is as follows: "*Our girls are growing up without any social marginalization: they are splendid, open minded, very kind, and have a great desire to learn and work"* (Prof. E. Sommavilla, WFL officer, personal communication, 2004).

This synthesis uses the same ingredients that allowed for the success of Mana's initiative, Mana has no illusions about the reality of the situation: "*the road towards eradication is still very long, especially for the pastoral nomads for whom the practice of infibulation has much deeper roots.*" We cannot forget that among the populations of the Lower Scebelli, an attenuated form of excision existed prior to recent events, allowing for the social prerequisites to acceptance of Mana's proposal.

References

1. Biasutti R. Razze e Popoli della Terra. Vol. 3. UTET, 1959. Torino.
2. WHO (World Health Organization) Female Genital Mutilation. Report of a WHO Technical Working Group, 17–19 July 1995, Geneva, 1996.
3. Martinelli M. FGM. Focus on Djibouti. Ph.D. Thesis dissertation, University of Padua: Rapporto sulla Povertà/Sviluppo Umano. In press, 2005.
4. Gasparrini S. Dijbouti fer de lance de la lutte contre les Mutilations Genitales. La Nation, 16/1/2005.
5. Grassivaro Gallo P., Livio M., Viviani F. Changes in infibulation practice in East Africa. Comments on a ritual alternative to infibulation in Merka, Somalia. pp. 133–142. In: Denniston G., Hodges F.M., Milos M.F. (eds). Flesh and Blood. Perspectives on the Problem of Circumcision in Contemporary Society. Kluwer Academic/Plenum, New York, 2004.
6. Zecchinelli C. No alle Mutilazioni Genitali, vince il fronte delle donne. Corriere della Sera, 4/2/2005.
7. NCTPE (National Committee on Traditional Practices of Ethiopia). Old Beyond Imagining Ethiopia Harmful Traditional Practices. Addis Ababa, 2003.
8. Piga De Carolis A. Il quadro etnico tradizionale nelle prospettive di sviluppo della Valle del Giuba. "Africa", 35:17–42, 1980.
9. Lewis J.M. Peoples of Horn of Africa. International African Institute. London, 1955.
10. Grassivaro Gallo P. La Circoncisione Femminile in Somalia. Una ricerca sul campo. F. Angeli (ed). Milano, 1986.
11. Grassivaro Gallo P., Rabuffetti L., Viviani F. Sunna gudnin. An alternative ritual to infibulation in Merka Somalia. pp. 99–111. In: Denniston G., Hodges F.M., Milos M.F. (eds). Understanding Circumcision. A Multi-disciplinary Approach to a Multi-dimensional Problem. Kluwer Academic/Plenum, New York, 2001.
12. Grassivaro Gallo P. Considerazioni su un'alternativa rituale all'infibulazione (Merka, Somalia). pp. 106–119. In: Scalari P., Ferruzza E. (eds). Dal confronto intergenerazionale allo sviluppo dell'identità del genitore. Atti dei seminari Gruppo "Cultura e Università" sett/nov. 2001, Ordine degli Psicologi, Consiglio Reg. del Veneto, Padova, 2002.
13. Gruenbaum E. Sexual response and Sudanese conceptions of the infibulated and uninfibulated vulva: Research in Sudan, 2004. Conference Advancing Knowledge of Psycho-Sexual Effects of FCM/C: Assessing the Evidence. Population Council, Egypt. Venue: Le Metropole, Alexandria, October 10–12, 2004.
14. Mackie G. Femal Genital Cutting: the beginning of the end, pp. 253–282. In: Shell Duncan B., Hernlund Y. (eds.) Female Circumcision in Africa. Culture, Controversy and Change. Lynn Rienner, Boulder, CO, 2000.
15. Various Authors. Les femmes de Djibouti font plier les religieux. La Nation, 8/2/05.
16. Various Authors Old Beyond Imaginings. Ethiopia Harmful Traditional Practices.

The Ritual Use of Herbs for Female Genital Modifications (FGMo) in Africa

Pia Grassivaro Gallo[1], Miriam Manganoni[1], and Franco Viviani[1,2]

Abstract Different herbs are commonly used in Africa during traditional practices connected to female genital modifications (FGMo) — expansive (genital stretching) and reductive forms (infibulation). The herbal component is an integral part of ritual genital stretching, often carried out in the grassy ground. Usually, these herbs act as lubricants and/or anesthetics to help labial manipulation. In the reductive forms of FGMo (i.e., infibulation), post-intervention herbal compresses, with hemostatic and cicatrizing functions, are placed on the wound. In particular, in the infibulation ritual, substances emitting marked aromatic perfumes are used. In Somalia, in the *ùnsi* ceremony, incense and myrrh are burned; in Sudan, in the *dukhàn* ceremony, sandal and acacia woods. These are the specific smoke ceremonies purifying women. In both expansive and reductive FGMo, a deep knowledge about the local herbs is reported, which connects the present African populations with those of the past, from which they probably inherited the knowledge and the utilization of FGMo interventions.

Introduction

During the Seventh World Congress on Sexuality (WAS, July 2005, Montreal, Canada) the interest of some researchers[1] was focused on the meaning of vaginal practices carried out in many Asian and African cultures. Preliminary results highlight that, beyond the several motivations for the practice (to increase the partner's pleasure, the vaginal "fashion," to eliminate unpleasant odors, and so on), the essence of all these initiatives appears to be sexually-linked. In other words, they contribute to build sexuality under the cultural profile. In our opinion, among these vaginal practices, female genital modifications (FGMo) assume a particular importance, and they are commonly considered an indispensable premise to marriage. Therefore, a survey aiming to understand the importance of the naturalistic aspect

[1]Padua Working Group on FGMo, Department of General Psychology, University of Padua, Italy
[2]Faculty of Medicine, University of Udine, Italy

G. Denniston et al. (eds.), *Circumcision and Human Rights*,

in genital modifications, carried out through analysis of the ritual herbs used in these practices, was considered to be a matter of concern.

Female Genital Modifications (FGMo)

WHO defines female genital mutilations (FGM) as all the procedures involving the partial/total removal and/or injury to the external genitalia, whether for cultural or other non-therapeutic reasons.[2] On the basis of the morbidity of the physical consequences induced by the latter, they are classified into three main categories (those most studied, ranging from Type I, removal of clitoris and/or clitoral hood; Type II removal of clitoris and labia minora and majora; and, Type III, removal of clitoris, labia minora and majoria, and stitching the vaginal opening), followed by a fourth one, which includes all other forms falling under the definition.[3] This valid effort, however, needs to be updated for two main reasons. First, all excised African women don't consider themselves to be mutilated. They consider the various alternative expressions used, i.e., FGM/cutting,[4,5] to be derogatory, making immigrant women feel denigrated (Marian Ismail, personal communication, 2001).[6] Secondly, the term mutilation does not include all the interventions on genitalia (i.e., the "expansive forms"). To find more appropriate terminology, we refer to the anthropologists of the past (Magigot, 1885, quoted in Puccioni, 1904[7]), who talked about "modification," an all-inclusive term without negative semantic connotations, also used recently.[6,8] As far as classification is concerned, we believe that a main distinction must be made with the same importance among the interventions, reductive and expansive, according to the table.[6] All interventions are collected into one global vision, as there is a common feature connecting them. They are seen as the pre-requisite to marriage and to the subsequent fertility of women.[9] In the table, the forms under examination in this text are shown in block letters.

Classification of Female Genital Modifications (FGMo)

Reductive forms	Expansive forms	Other
Excision (clitoris, *labia*)	**Stretching (clitoris, *labia minora*)**	Ritual defloration (hymen)
Vaginal restrictions, **infibulation** (perineal)	Dilation (of the vaginal channel, of the vagina)	Elimination of the natural lubrication through the insertion of grass, ground stones, etc.
Castration, sterilization (uterus)		Introcision Abortion practices

However, a dichotomy exists between the modalities in which FGMo are carried out (reductive and expansive interventions), which is confirmed by a different geographic distribution, a specific set and use of terms regarding the sexual sphere, the utilization of vegetal substances accompanying the interventions, and the ecology of the population carrying out genital alterations.[11]

Materials and Methods

To investigate the role of plants in FGMo, we utilized:

1. Different vegetal samples collected during two field surveys carried out in Africa — in Uganda (2002) and in Malawi (2004)
2. Photos of plants photographed *in loco*
3. The graphic representation of central-African adolescents asked to draw circumcision
4. Information from interviews with local informants in Uganda and Malawi, Somalia, and Sudan
5. Local references
6. Analytical research (reviews of historical, anthropological, and ethnographical data)

(1, 2) In particular, the scientific ascertainment of vegetal species utilized by traditional performers in reductive and expansive FGMo was carried out by specific departments of the University of Padua, (Prof. P. Giulini) from the Department of Biology and Pharmacy; for vegetal samples the utilized nomenclature refers to the Index Kewensis (1997),[10] consulted by the Royal Botanic Gardens, Kew (London). On the whole, some tens of vegetal samples are represented, many classified from the botanic viewpoint, shown in two tables built respectively for the expansive FGMo (Table 1) and the reductive forms (Table 2).

Table 1 Herbs used in expansive FGMo forms in the great African lakes region (Uganda, Malawi, Congo, Mozambique)

		Scientific determination		
Source	Local name	Family	Gender	Species
The Great Lakes Region		*Fabaceae*	*Erythrina* L.	*abyssinica*
Kashamura (1973)[15]				
Uganda				
Weimberg et al. (2004)[17]	*Colocynth berries*	*Cucurbitaceae*	*Citrullus* L.	*colocynthis*
	Kabbo ka bakyala			
	Kanyebwa	*Oxalidaceae*	*Oxalis* L.	cfr. *latifolia*
Villa (2002/2003)[16]	*Kajampuni*	*Oxalidaceae*	*Oxalis* L.	
	Mukasa			
	Namirembe			
	Ntengotengo	*Solanaceae*	*Solanum* (Tourn.) L.	*incanum*
Malawi				
Moro (2004/2005)[12]	**Chisio*			
*Ottarda M. pers. comm. (2005)	**Eleusine, Mawe*	*Graminaceae*	*Eleusine* Gaertn.	*coracana*
	**Kambalimozi*			
	Khapale	*Fabaceae*	*Julbernardia*	*globiflora*

(continued)

Table 1 (continued)

Source	Local name	Scientific determination		
		Family	Gender	Species
	**Mwuleme*		Pelleg.	
	Mwunguti	*Bignoniaceae*	*Kigelia* DC.	*africana*
	**Nalitso*			
	Napini	*Combretaceae*	*Terminalia* L.	*sericea*
	Nkulitsa	*Leguminosae*		
	Nsatsi	*Euphorbiaceae*	*Ricinus* L.	*communis*
	Nthula	*Solanaceae*	*Solanum* (Tourn.) L.	*aculeastrum*
	Nyidi	*Hydrocharitaceae*	*Ottelia* Pers.	*ulvaefolia*
		Rhamnaceae	*Ziziphus* Mill.	*jujuba*
Congo				
Evans-Pritchard (1937)[18]	*Eleusine Nganza*	*Graminaceae*	*Eleusine* Gaertn.	*coracana*
Mozambique				
Junod (1912)[19]		*Solanaceae*	*Datura* L.	*fastuosa*

Table 2 Herbs used in reductive FGMo forms in Ethiopia, Somalia, Sudan

Source	Local name	Scientific determination		
		Family	Gender	Species
Ethiopia				
Huber (1966)[25]	*Kosso, Cusso*	*Rosaceae*	*Brayera* Kunth	*antihelmintica*
Vergiat (1951)[35]		*Vitaceae*	*Ampelocissus* Planch.	*abyssinica*
NCTPE (2003)[48]		*Brassicaceae*	*Lepidium* L.	*sativum*
Somalia				
Grassivaro and Vivani (1992)[39]	Incense *Dir, Lubàn, Fooh,'Dihdîgu*	*Burseraceae*	*Boswelia* Roxb. Ex. Colebr.	*sacra*
	Mirra	*Burseraceae*	*Commiphora* Jacq.	*abyssinica*
Erlich (1986) °De Villenuve (1937)[28]; Pieters (1972)[29]; David (1978)[30]	(*Mal-mal:* mixture of substances, among them myrrh)	*Mimosaceae*	*Acacia* sp.	
Sudan				
Lightfoot-Klein (1991)[43]		*Santalaceae*	*Santalum* L.	*Album*
Bedri 1993 in Lovel et al. (2004)[44]		*Mimosaceae*	*Acacia* L.	*Nilotica*

(3) During field surveys on FGMo in Uganda, Malawi, and Somalia, some graphic methods were administered to the examined girls, i.e., adolescents were requested to draw what they remembered about female circumcision in their region.[11–13] From this material, some examples were extracted that refer to the subjects' recall of the choice of ritual plants, carried out by the operator before circumcision, or they refer to the naturalistic frame in which the rite was performed. Some of the drawings are presented here (Figs. 1–6). From the graphic material collected in Somalia, no naturalistic information was drawn because circumcision is often carried out privately and the landscape is generally arid. (4) During the interviews with operators, they often showed how the plants were treated (burned, powdered, roasted, boiled, and so on) and which parts (leaves, roots, fruits, branches, and so on) were used for ritual purposes.

Fig. 1 Self-manipulation in the bush[12]

Fig. 2 Self-manipulation girl with a friend. Daily elements and vegetation[12]

Fig. 3 *Ssenga* and girl in the bush with ritual herbs[16]

Fig. 4 Seeds facilitating genital stretching[16]

Results

Vegetables and Expansive FGMo Forms

The naturalistic component is an integral part of the specific ritual of labial stretching, which in some countries is carried out in the countryside, considered a sacred place. In Uganda, for example, the rite is called "visiting the bush or the forest"[14]; while the Great Lakes populations[15] used the French phrase, "bois sacré." Even when the rite is performed privately (hut, house, and so on), the operator generally uses different plant extracts for genital manipulation.

Often, a clear symbolic meaning oozes from vernacularisms, alluding to the sexual sphere and/or to fertility. *Nyini*, a plant used in Malawi, refers to the female sexual organ).[12] Among the Ugandan Baganda, *Namirembe* means "the one who gives peace," and this is a good wish for marital serenity and stability. *Mukasa* is the name of both Lake Victoria and the productive fertility living in it. The expression, *Kabbo ka bakyala*, literally means "cup" or "the women's basket."[16]

Fig. 5 Genital stretching in the bush[16]

Fig. 6 Genital stretching in the banana plantation[16]

Photo 1 *Oxalis* cfr.*latifolia* H.B. et K. (Villa, 2002/2003)

In Table 1, herbs used in the expansive FGMo forms of the Great Lakes area are listed,[15] in particular, for Uganda,[16,17] Malawi,[12] Congo,[18] and Mozambique.[19]

For Uganda, the collected vegetal samples that were possible to scientifically ascertain, based on traditional operators' descriptions, refer to *Citrullus* L., *Oxalis* L. and *Solanum* L. For the others, only the local name is known (Photo 1).

For Malawi, the list is more ample and differentiated, containing different families: *combretaceas, leguminosae, euphorbiaceas, hydrocharitaceas, bignoniaceas, and solanaceas*.

To complete the descriptive set, some graphic representations of a group of Malawian (Figs.1 and 2) and Ugandan (Figs. 3–6) adolescents are reported, which portray the natural environments and the choice/collection of plants useful for the labial stretching ritual.

Vegetables and Reductive FGMo Forms

Specific plants are used even among populations "reducing" genitals (exemplified by those who practice infibulation in the Horn of Africa) (Table 2).

During the post-operation phase of infibulation, we mention:

- Resorting to the use of hemostatic plasters, such as butter and tree bark-based in Kordofan[20]; flour-based among the *Kabâbish*[21]; and, among others, ash-based[22] among the *Leyla* of Higher Volta,[23] raw eggs and *hennè*[24] in Nubia, hot butter, flowers, and leaves of *Kosso* (*brayera anthelmintica* Kunth) among the *Amhara* of Etiopia,[25] boiling oil,[26] those made of a resinous coating[27] among the *Afar* (Ethiopia), and of an adhesive mixture based on myrrh dust, sugar, lemon juice, egg yolk, and rubber (*mal-mal*) among the *Somalis*.[28–30]

- Utilizing fomentations of vegetal decoctions (leaves and barks) to block hemorrhage among the *Jekri* and the *Sobo* of Nigeria[31]; *neb neb* (*Acacia*), gypsum and barks of *koloko* among the *Barbara*[32]; crushed leaves or banana sap among the *Soussou*[33]; leaves placed on fire among the *Coniagui*[34]; *Ampelocissus* poultices among the *Manja*[35]; and a purée of *ngar* (mucilage) among the *Sara*.[36]
- Other vegetables, singularly used for rituals in Ethiopia, Somalia, and Sudan, are listed in Table 2 as well.
- Some plants (such as incense, myrrh and sandal), even if different from each other botanically, are pooled from a particular characteristic — during brisk combustion they release a marked aromatic olfactory scent. The traditional use of these contributes, in the "smoke ceremony,"[37,38] to purify the female body (before sexual intercourse or after parturition) and allows renewal of the sexual relationship with the partner. Furthermore, such aromas substitute individual olfactory scents which, in the case of infibulation, are reduced because of the narrowing of the vaginal channel.[39]

In Somalia, the ritual herbs found to be more important from this point of view are listed below (in the north-eastern part of the country, less valuable species were signalled)[40]:

Boswelia sacra (Birdw.) Flueck (incense tree) (Cavestro, personal communication, 2004), whose vernacular names are *dir*, *lubàn*, and *fooh*.[41]

Commiphora abyssinica (myrrh) is normally powdered and used with natural incense and other vegetable substances to purify the female body in the *ùnsi* or the "smoke ceremony."[38,41] The woman undertakes *ùnsi* before a sexual relationship or after labor, allowing the partner to resume sexual relationships. Moreover, myrrh is used during the post-infibulation phase *mal-mal* (prepared as already mentioned).

Santalum album (sandal wood) and other resinous woods are burned there and in the Sudan during the *dukhàn*, the "smoke ceremony." The most common use for infibulation, however, is obtained from the stipels of the acacia tree — *Acacia nilotica* (L.) Wild ex Delile — that, in the traditional way, are fixed in an alternate manner in both the operated great lips; then, among the protruding heads of the same stipels, an animal twist is passed, to permit the margins of the sewing to adhere. To promote a quick scar-tissue healing, *mal-mal* is applied.[37]

The "Smoke Ceremony" and the Siccative Fumigations

As far as Sudan is concerned,[42,43] describes, among the purification rites following episiotomy (the cutting carried out to permit delivery in infibulated women), the "smoke ceremony" or *dukhàn*. The same ceremony, also from Bedri, was described in 1993 (reported in Lovel et al., 2004[44]), as being carried out with siccative fumigations with aromatic substances (incense and myrrh) in order to help the healing process. This ritual is so common throughout the country that the scent of the braziers pervades roads, shops, and markets, allowing identification of the smell as being "the same scent of the woman."[42]

This ceremony, characteristic of *post-partum,* is practiced one or two times per day, till the end of the puerperal period. Small pieces (around 15 cm) of sandalwood are prepared, then put into a 60 cm deep hole and later on burned with charcoal. The woman, undressed, rubs her body with oil and then sits on the floor with her legs across the hole and covers herself with a blanket. The fumes, absorbed by her oily body, open the pores of the skin and facilitate the purification process.

In northern Somalia, the smoke ceremony is called *ùnsi*.[41] The woman, in a brazier, burns a particular type of incense (*dir*), available in markets, together with other fragrances collected in the bush. The burning of these substances emanates a very thick and aromatic smoke. In the meanwhile, the woman washes herself and, from this moment on, she will avoid all heavy jobs to prevent sweating. She puts on clean frocks, the newest she owns, and she crouches down close to the brazier, letting the smell permeate her body. Later on, she smears her body with an oil (*labèn*), pulled out from a tray that she keeps in her house, which is composed of fat that grows up on the surface of yogurt during fermentation or on top of milk, in order to let the smell of the burned substances be fixed to her body and hair.

Always in Somalia, in the *ùnsi* ceremony,[38] the incense (*lubàn*) assumes the meaning of a sexual signal; from this perfume, in fact, the husband understands the partner's willingness to undertake sexual intercourse, a readiness that will be considered improper if expressed verbally.

More recently, Mana Abdurahman Ali Issa (personal communication, 1998 and 2005) specifies that, for this ritual, incense is not utilized anymore — a blend of vegetal aromas, perfumes, and sugar is used. In the market, it is possible to buy some of these ingredients in bags that will be smashed and crushed in a mortar, and placed on a fire with a kilogram of sugar. In this way, a paste, overabundant for personal use, will be obtained, and it could be exchanged with other women or sold in the market later on. After intercourse, it is smeared on the body and on the hair, previously wet to fix the smell. Furthermore, it appears that the generalized use of *ùnsi* is strictly connected to the infibulatory practice and to the connected problems. In fact, the aim of the ceremony is to avoid unpleasant smells from urinary losses, coagulation of menstrual flux, thwarted in its outflow, and, in general, from possible fistulas caused by the intervention.

A practice of siccative fumigations utilized after infibulation in Somalia has been already quoted by Costanzo in 1968.[45] In particular, as combustible materials vary, dry woods derived from a shrubby plant (*ghedbe*) are used. They are put in a hole in the soil (Fig. 7). They waft, burning slowly, an aromatic smoke passing from the combusting hole to the hole upon which the girl is sitting. But if there is only one hole, the infibulated girl sits on it, and soon the fire is extinguished. The steaming hot ash would purify and, therefore, will furnish an antiseptic protection to the wounded anatomical structure, which otherwise will be exposed to different germs;

Fig. 7 Cicatrizing fumigation for infibulation "sewing" in Somalia[45]

the other aim is to soften the part. Such a practice, lasting 10–15 min, is repeated daily for at least a week after the intervention.

Active Ingredients and Pharmacological Functions of Analyzed Vegetables

The ritual herbs, classified botanically in the expansive FGMo forms, substantially act as lubricants (*Ricinus communis* L.) and local anaesthetic (*Erythrina abyssinica* L., *Khosomole, Mukasa, Nkulitsa*), as they facilitate labial manipulation. Apropos, we remember some *Baganda* women who immigrated to Rome and unsuccessfully attempted to grow plants from Uganda, They substituted cold creams for ritual purposes (Photos 2 and 3).[46,47]

Other essences directly act on female genitalia in a chafing way, directly determining intumescence and soreness of the perineal region, such as *Citrullus colocynthis* (L.) Schrad[19] and extracts of the root (*Nyika, nyendere*) of the *Ottelia ulvaefolia* Walp.

Other plants can modify the vaginal orifice, narrowing it (*Chisio, Terminalia sericea Burc* ex DC.) or widening it (*Kigelia africana Benth.*), or they are used for ritual defloration (*Ziziphus jujubalam*).

In FGMo reductive forms, we highlighted the following connection between vegetables and infibulation — the use of concoctions, pastes smeared on scars, and siccative fumigations, having hemostatic and cicatrizing functions. The "smoke ceremony," common in all of East Africa, where it is carried out with different aromatic substances of vegetal origin, fulfils a three-fold function during confinement: to dry, to scar over, and to purify external genitalia,[38] plus the function of providing a sexual signal indicating a desire for intercourse.[41]

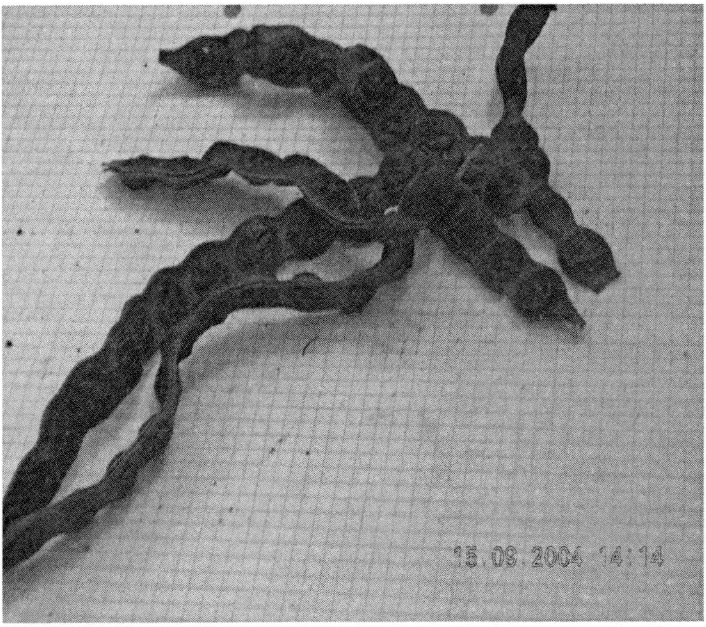

Photo 2 *Nkulitsa* (Moro, 2004/2005)

Photo 3 *Ricinus communis* L. (Moro, 2004/2005)

Comment and Conclusion

The noteworthy importance of vegetables in the rituals of FGMo, often noted in the drawings of the adolescents tested, was highlighted in the analysis. In fact, both the frame of the rite and the herbal choice, led by the operator, were drawn.

In general, the local populations acquired only empirical information on the therapeutic value of the substances they ritually use; such knowledge, orally passed down from generation to generation, constitutes the memory of the ethnic group. Conversely, no knowledge of active ingredients contained in the plants was ascertained.

The impression, from the pharmacologic analysis carried out (compared with the information furnished by local informants), is that the local traditional knowledge is rather general and that the plants are used in a specific way, as they are utilized for different pathologies, dissimilar to those ritually-connected to FGMo.

From close inspection of the listed plants (Tables 1 and 2), some botanic heterogeneity is derived. Only the gender *Solanum* is utilized both in Uganda and in Malawi and the gender *Eleusine* is used both in Malawi and in Congo. Their utilization is in accordance with the expansive FGMo culture. They act as lubricating substances favoring manipulation for stretching genitalia or as irritating substances, swelling the anatomical structures to be kneaded (*labia minora* and clitoris).

Conversely, the function of vegetables in the reductive FGMo forms is neatly differentiated and specific. Their pharmacologic function, as already mentioned, is to dry, cicatrize, and purify the wound following infibulation (Photos 4–8).

In our opinion, the most interesting aspect is the use of aromatic plants in the fumigation ceremonies, which appears to be common in all of east Africa and, therefore, specific to the infibulatory populations. From these rituals, the importance of the olfactory aspects is inferred. In the past, this was interpreted by us as being a substitute of the vaginal emanations that were stifled by the vaginal narrowing following infibulation.[39] Furthermore, it has the value of a sexual signal, replacing the explicit language in societies that humble their sexuality instead of magnifying it, as happens among the expanders.

Photo 4 *Solanum aculeastrum* Dun. (Moro, 2004/2005)

Photo 5 *Ziziphus jujuba* Lam. (Moro, 2004/2005)

Photo 6 *Boswelia sacra* (Birdw.) Flueck

Photo 7 *Commiphora abyssinica* Engl.

Photo 8 *Acacia nilotica* (L.) Wild ex Delile

In conclusion, even the analysis of the vegetable aspect highlights the duality of FGMo processes. However, another aspect emerges — the deep knowledge of the local naturalistic aspects, and this appears to be an important *trait-d'union* with the populations of the past, from which African people today have probably inherited FGMo interventions.[12]

References

1. Hull, T., Martin-Hilber, A., Utomo, I., Im-Em, W., Widyantoro, N., Preston-White, E., Bagnol, B., Smit, J., Vaginal Practices in Asia and Africa. Paper presented at the 7° WAS, 10–14/07/05, Montreal, 2005.
2. WHO, Female Genital Mutilation. Report of a WHO Technical Working Group, 17–19 July, 1995. Geneva, 1996.
3. UNICEF, Report on UNICEF Global Consultation on Indicators: Female Genital Mutilation and Cutting (FGMC) and Child Marriage. New York (unpublished), 2003.
4. Shell-Duncan, B., Hernlund, Y., Female Circumcision in Africa: Dimension of the Practice and Debates, pp. 1–39. In: Shell-Duncan B. & Hernlund Y. (eds.) Female "Circumcision" in Africa Culture, Controversy and Change. Lynne Rienner, Boulder, CO, 2000.
5. Gruembaum, E., The Female Circumcision Controversy. An Anthropological Perspective. Pennsylvania University of Pennsylvania Press, Philadelphia, 2001.
6. Fusaschi, M., I segni del corpo. Per un'antropologia delle modificazioni dei genitali femminili. Bollati Boringhieri, Turin, 2003.
7. Puccioni, N., Delle deformazioni e mutilazioni artificiali etniche più in uso, Archivio per l'Antropologia e l'Etnologia, 34: 391–401, 1904.
8. Erlich, M., La Femme Blessée. Essai sur les mutilations sexuelles féminines. L'Harmattan, Paris, 1986.
9. De Rachewiltz, B., Eros Nero. Costumi Sessuali in Africa dalla Preistoria ad Oggi. Longanesi, Milan, 1963.
11. Index Kewensis 2.0 cd-rom. Oxford University Press, Oxford, 1997.
12. Grassivaro Gallo, P., Tita, E., Viviani, F., At the Roots of Female Genital Modifications (FGMo), pp. 159–169. In: Bodszar E. B. & Susanne C. (eds.) Human Evolution: Facts and Factors, Bienniel Books of European Anthropological Association, Vol. 4. Eötvös University Press, Budapest, 2006.
13. Moro, D., Modificazioni Genitali Femminili in Malawi. Resoconto di un'indagine sul campo. Tesi di Laurea in Psicologia, University of Padua, Italy, Academic Year 2004/2005.
14. Moro, D., Grassivaro Gallo, P., Passaquieti, A., Female Genital Modification in Malawi: A Field Research, page 9. The 9th International Symposium on Circumcision, Genital Integrity, and Human Rights. Program and Syllabus of Abstracts, University of Washington, Seattle, 2006.
15. Sengendo, J., A Cultural Approach to HIV/AIDS. Prevention and Care. UNESCO/UNAIDS, Research Report in Uganda's Experience. Country Report Studies and Reports, Special Series, Issue 1, Cultural Polices for Development Unit, UNESCO, 1999.
16. Kashamura, A., Famille, Sexualité et Culture. Essai sur les moeurs sexuelles et les cultures des peuples des Grands Lacs Africains. Payot, Paris, 1973.
17. Villa, E., Longininfismo rituale tra le Baganda (Uganda). Tesi di Laurea in Psicologia, University of Padua, Academic Year 2002/2003.
18. Weimberg, L., Mayer Fuchs, C., Brooke-Condom, J., Leigh Hobbs, J., Ortiz, A., Supplemental pretrial brief in support of respondents' application for asylum, withholding of removal, and withholding of removal under the convention against torture (unpublished data). Documents for the Assistant District Counsel. Boston (MA): U.S. Department of Justice Executive Office of Immigration Review, Immigration Court, October 26th, 2004.
19. Evans-Pritchard, E. E., Witchcraft, Oracles and Magic among the Azande. Clarendon, Oxford, 1937.
20. Junod, H. I., The Life of a South African Tribe. University Books, Hyde Park, New York, 1912.
21. Pallme, P., Travels in Kordofan. Hadden, London, 1844.
22. Seligman, C.G., Seligman, B.Z., The Kabâbish, a Sudan Arab Tribe, pp. 105–184. African Studies, II, Harvard University Press, 1918.

23. Fröhlich, W.G., Sitten und Gebruchäuche der Nubier insbesondere die Mädchenbeschneidung und ihre Folgen, pp.791–796. Schweiz Medizinische Wochenschrifts, 1921.
24. Driot-Nisenmann, M.H., L'excision génitale chez l'Africaine et ses conséquenes médicales. Thése de Medicine, Paris, Saint-Louis-Lariboisière, 1979.
25. Kennedy, J.G., Circumcision and excision in Egyptian Nubia. Man, 5 (2): 175–191, 1970.
26. Huber, A., Female Circumcision and Infibulation in Ethiopia, pp.87–91. Acta Tropica, Basel, 23 (1), 1966.
27. Creignou, Dr., L'infibulation en pays Somali, pp.111–117. Institut Français d'Anthropologie, Paris, 1912.
28. Albospeyre, M. Les Danakil du Cercle du Tadjura, pp. 103–161. Cahiers de l'Afrique et de l'Asie, 5°, Paris, 1959.
29. De Villeneuve A., Etude sur une costume somalie: les femmes cousues. Journal de la Sociètè Africaine., 7: 15–32, 1937.
30. Pieters, G., Gynécologie au pays des femmes cousues. Acta Chirurgica. Belgica, 3: 173–193, 1972.
31. David, A., L'infibulation en République de Djibouti. Thèse de Médecine n°131, Université de Bordeaux II, 1978.
32. Granville, R.K., Notes on the Jekris, Sobos and Ijos of Wari district on Nigeria Coast Protectorate, Vol. I, pp.104–126. Journal of the Royal Anthropological Institute, New Series, 1899.
33. Traoré, D., Piquére de scorpion, pp.423–424. In: Médecine et magie africaines ou comment le noir se soigne-t-il? Présence Africaine, Paris-Dakar, 1965.
34. Talbot, D., Circoncision rituelle de la femme dans l'Afrique occidentale. Annales d'Hygiene et Médicine Coloniales, 3: 283–286, 1900.
35. De Lestrange M., Les Coniagui et les Bassari. P.U.F., Paris, 1955.
36. Vergiat, A.M., Les rites secrets des primitifs de l'Oubangui, Payot, Paris, 1951.
37. Muraz, G., Satyres illustrées de l'Afrique noire, du Comité de Documentation et de Propagande de l'Afrique noire français. Paris, 1947.
38. Grassivaro Gallo, P., La circoncisione femminile in Somalia. Una ricerca sul campo. F. Angeli, Milan, 1986.
39. Diasio N., Il vaso di Pandora, ovvero della chiusura del corpo femminile. Aspetti storico-antropologici, pp. 23–53. In: Mazzetti M. (ed.) Senza le ali. Le MGF, F. Angeli, Milan, 2000.
40. Grassivaro Gallo P., Viviani F., The origin of infibulation in Somalia: An ethological hypothesis. Ethology and Sociobiology, 13: 253–265, 1992.
41. Senni, L., Gli alberi e le formazioni legnose della Somalia, pp.84–101. Istituto di Agricoltura Coloniale, Florence, 1935.
42. Grassivaro Gallo, P., Figlie d'Africa mutilate. Indagini epidemiologiche sull'escissione in Italia. L'Harmattan Italia, Turin, 1998.
43. Lightfoot-Klein, H., Rites of purification and their effects: some psychological aspects of female genital circumcision and infibulation (pharaonic circumcision) in an Afro-Arab Islamic society (Sudan). Journal of Psychology & Human Sexuality, 2: 79–91, 1989.
44. Lightfoot-Klein, H., Orgasm in Ritually Circumcised African Women. First International Conference on Orgasm. New Delhi, February 3–6, 1991.
45. Lovel, H., Bedri, N., Zeinab Mohamed, Njikam, Y., Rhetta Moran, Mc Gettigan, C., Efm Sa'ad, Khaled, M.A., Abdelmagied, A.M., Efua Dorkenoo, The Practice of Re-infibulation,Findings of a Systematic Search of English and Arabic Literature. Seminar of Advancing Knowledge of Psycho-Sexual Effects of FGM/C: Assessing the Evidence. Alexandria, Egypt, October 10–12, 2004.
46. Costanzo, G. La circoncisione femminile in Africa. 3° Convegno Nazionale di Antropologia Culturale – Perugia 1968. Risa, Rome, 1968.
47. Grassivaro Gallo, P., Villa E., Longininfismo rituale tra le Baganda (Uganda). Rapporto Preliminare. Rivista di Sessuologia, 28: 17–22, 2004.
48. Grassivaro Gallo, P., Villa, E., Ritual Labia Minora Elongation among the Baganda Women of Uganda, Psychopathologie Africaine, 2006, in press.

49. NCTPE (National Committee on Traditional Practices of Ethiopia), Old Beyond Imagining Ethiopia Harmful Traditional Practices, Addis Ababa, 2003.
50. Grassivaro Gallo, P., Moro, D., Modificazioni Genitali Femminili in Malawi. Primo resoconto di un'indagine sul campo, pp. 87–102. In: Grassivaro Gallo P. & Manganoni M. (eds.) Pratiche tradizionali nocive alla salute delle donne. Unipress, Padua, 2006a.

Female Genital Modifications in Malawi
Culture, Health, and Sexuality

Pia Grassivaro Gallo, Debora Moro, and Miriam Manganoni

Abstract A strong ethnic heterogeneity has made difficult the synthesis of the results of the fieldwork in Malawi (Mangochi district), organized by the Working Group on the FGM of the University of Padua in 2004 and focused on the analysis of FGM. The most important data deal with labial stretching, practiced systematically on the Malawi teenagers during the segregation (*chiputu*), lasting a few weeks in a hut under the guide of a female instructor (*namkungwi*) and the (*phungu*) responsibility of the grandmother. Also the psycho-physical connections and reflections on sexuality are presented.

Introduction: The Fieldwork

Malawi population, strongly hybridized, is composed of extremely heterogeneous Bantu groups — Alomwe, Achewa, Ayao, Angoni, Anyanja, Atumbuka, Atonga, and Sena. Some migrated here from South Africa, others from Mozambique.[1] They have different traditions, ways of life, languages, customs, and religions, yet female genital modification is a common trait to these peoples, as well as the Great Lakes peoples.[2] It has different aspects within specific rituals.

The mixing of different ethnic groups inside the same village has also interfered with the organization and the performance of the rituals themselves, therefore, it is difficult to point out a cultural behavior valid for the entire country. In addition, the smaller groups have the most important cultural behavior, due either to social enhancement or lack of economic resources allowing autonomous celebrations.

During the summer of 2004, the Working Group on FGM of the University of Padua organized fieldwork in Malawi, with the collaboration of two Monfortan missionaries from Redona (Bergamo), Father Luciano Nervi and Father Angelo Assolari. Moreover, in the fieldwork, we collaborated with Monseigbeur Joseph Kimu (of the Yao ethnic group), Dean of the St. John's Major Seminary; he also introduced us to a young Yao woman interpreter (F.M.); Father J. C. Chakanza,

Padua Working Group on FGM, Department of General Psychology, University of Padua, Italy

Chancellor College professor at the University of Zomba, provided precious bibliographic resources. The fieldwork took place in the Mangochi district, comprised of the town of the same name and the neighboring villages in the south of the country, between the lakes of Malawi and Malombe.[3] During the fieldwork, anthropological/naturalistic and psychological/sociological information was gathered.

Methodology and Subjects Surveyed

The following tools were used for the analysis of the cultural aspects concerning FGM in Malawi:

(a) Drawing[4] of some elementary teachers and of 95 female students (10–20-years old). They drew what they remembered or knew, depending on whether they had been either submitted to the rite (*chiputu*) or not, and how the manipulated and non-manipulated women appeared. Beneath the drawings, the girls wrote their observations. The subjects, of both rural and urban extraction, were met in the schools of the Mangochi district, St. Monica and St. John, in the urban schools, and in the villages of Nsanyla and Chopola.
(b) Interviews with seven wise women, of whom two were instructors in the chiputu and, therefore, ritual operators, and with two men, all from the Magochi District and belonging to the Lomwe, Ngoni, Yao, Chewa, and Sena.
(c) Photos of the perineal region of two old manipulated Lomwe women and of one during an auto-manipulation.
(d) Photos and information gathered during a ceremony in Nsanyra (Mangochi).
(e) Bibliographical data from local books, articles, and personal communications from the above Malawian referents and Monfortan Fathers.

As to the aspects concerning health and sexuality connected with labia minora stretching, we analyzed the contents of the interviews of the adults and the comments of the girls accompanying their graphic representations. This qualitative analysis focused on the most frequent motives, either positive or negative. These motives were grouped in Tables 1 and 2, dealing respectively with aspects of physical and mental health and sexuality. For comparison's sake, we added information from fieldwork in Uganda in 2002.[2,5,6] All the material was gathered either directly in English or in the Chichewa language and translated into English.

Results: FGM in Malawi

Genital manipulations in their complexity comprise the following interventions:

- Labia minora stretching (personal communication 2004)[7–11]
- Widening of the vaginal canal (personal communication 2004)[10]
- Ritual defloration, either with introduction of thorny plants (personal communication 2004) or at first intercourse with the *fisi*[8,9,12,13]

Table 1 Health aspects of ritual stretching

Advantages (either real or assumed)	Malawi	Uganda
Eliminates the girl's pain at first intercourse	X	X
Helps woman at parturition	X	X
Re-closes vagina after parturition	X	
Protects vagina from dust/germs	X	
Disadvantages		
Pain at first stretching	X	X
Pain if performed late	X	X
Encourages sexual promiscuity	X	
If the labia are too long:		
Obstructs fast walking	X	
Being seated is uncomfortable	X	
Vagina too lubricated	X	
Labia can seal	X	
Clotted blood favors bacteria	X	
Mental Health		
Advantage: stretching avoids peers' teasing	X	
Disadvantage; girl uncomfortable with non-manipulated peers	X	

Table 2 Ritual stretching and sexual aspects

Advantages (real or presumed)	Malawi	Uganda
Shows the woman is ready for penetration by men with different sized penises	X	
Increases men's pleasure, favors ejaculation and penetration	X	X
Men are more faithful to wives	X	
Increases women's pleasure during petting		
Beautifies vagina	X	
Women feel more beautiful and passionate	X	
Increases pleasure for both partners	X	X
Disadvantages		
Makes women feel different due to genital modification	X	
Is a task to be performed every day	X	
Some men may dislike stretched labia	X	
Disadvantages of not being stretched		
Intact women laughed at by girl friends	X	
More painful penetration due to tight vagina	X	
More painful intercourse for men	X	

- Introduction in the vagina of anesthetic herbs, either to alleviate pain during the first intercourse or reduce the dimension of the structure (personal communication 2004), pulverized stones, antiseptic herbs, products from the local pharmacopoeia, fabric, or paper in order to dry the normal vaginal lubrication[14]

These interventions last several years and accompany a girl while growing until marriage and the birth of the first child. They occur within specific rituals.

The interventions, of course, do not occur because of any sanitary indication, but because of cultural imperatives. According to the classification of the so-called Female Genital Mutilations of the WHO (1996),[15] they are included in the poorly defined fourth group, for exclusion from the other three groups. As a matter of fact, they are not real mutilations, but more precisely *manipulations*, mostly of an expansive nature. Thus, we prefer to adopt the term female genital modifications (FGMo)[2] to define them. In addition to these expansive types of FGMos, in Malawi a "reductive" FGM has been observed, i.e., clitoridectomy (Group 2),[15,16,22] which would occur among the Lomwe.[9]

As to the rituals we mentioned above, we distinguish, in particular:

- A collective segregation connected with the pre-pubertal rites of passage (*chiputu*), within which the systematic elongation and stretching of the labia minora occur (stretching). To this, the widening of the vaginal canal and other manipulations can be added. The segregation ends with the ritual defloration by means of a tool or a first intercourse with a specific person — the *fisi* (literally, the hyena). This ritual occurs when the girl is 8 to 13 years old and is completed with sexual instruction and teachings about the correct sexual behavior and, more generally, the future married life. The *chiputu* is concluded by a day of feasting, when the initiate girls, as "new women," are presented to the community.
- A second period of segregation, at menarche (*chinamwali*), when checking the genital manipulations and the completion of sexual instruction take place. This step is apparently the most important from a social point of view, and is completed by the performance of the *fisi*, if it has not occurred before. In Malawi, it must be noted, sexual maturation occurs at the age of 14.3 years: DS 1.4 years.[8]
- The ceremonies for the traditional wedding and those for the birth of the first child.

Labial Elongation

Culture

The data gathered in 2004 in Malawi dealt mostly with the *chiputu* ritual and thus labial stretching.

The main points are the following:

- The first intervention of genital manipulation takes place among girls, 7–9 years old, outdoors (in the bush, during the collective bathing, when they play family relations — *mansanje*) (Fig. 1). It begins with everyday contacts with elder women of the family or the village, who first only suggest, then begin to pressure as menarche nears. The invitation is always indirect, "*Go and play with your mates*." The manipulation itself is encouraged among all the ethnic groups, although its beginning seems to be a personal choice. The girls, however, live under strong familial and social pressure favoring the decision to start manipulation. In fact, the girl who decides not to submit herself to the ritual

Fig. 1 Automanipulation in the bush

lives in a state of indefinite immaturity, which marginalizes her socially. In practice, the ritual takes place this way: the girl extends, stretches with her hand, the external side of the labia minora, often softening it with products from specific vegetal essences. Always informally, the widening of the vaginal canal may occur concurrently, manually or through the introduction of vegetables (maize corncobs, manioc roots, etc),[10] thorny branches. The body position (suggested by expert mates and later perfected by a female instructor) of the girl at this stage is interesting: the girl squats with her arms reaching the vaginal area in an arc under her thighs. The manipulation is always drawn graphically this way. During the manipulation, the girl uses the first two or three fingers, and the time of the intervention is about 10–15 min, a time that decreases as the labia elongate.

- Systematic stretching is performed by and under the guidance of a female instructor (*anamkungwi* or *angolosolo*), chosen by the family, when the girl enters the pre-menarche age (between 8 and 13). The female instructor first checks the girl's genitals at home, in order to see whether the manipulation has been performed correctly; then, she operates in a systematic way for a group of girls in a hut built either in the village or in the woods (*tezo* or *simba*). The hut will be destroyed at the end of the rituals. The initiates will be secluded inside the hut for some weeks (*chiputu*), together with their female instructors and one or two godmothers (*phungu*) for each girl. The godmothers collaborate with the instructor and are responsible of the girls' behavior as well (Fig. 2).

In the meantime, the instructors transfer their specific cultural lore, which deals with the practical behavior during sexual intercourse, in order to please the partner, respect for the spouse and the elders, and understanding of mourning rituals and community ceremonies as well was everyday life. The initiates learn traditional songs and dances, and — above all — perfect the technique of labial elongation, performing it during the *chiputu*, also with the help of vegetal essences. All this is learned through the female instructors' rehearsals and pantomimes. The *chiputu* ends with the checking of a culturally adequate labial elongation (as long as the middle finger or a little more), and some sporadic perfection of other FGMos, such as the widening of the vaginal canal and ritual defloration.

Fig. 2 Two subjects in automanipulation

The "new women" will be presented to the village community with a day of feasting. The girls have their shaved heads smeared with a mud cap, their bodies ornamented with many strands of beads along the neck, wrists, and back. Dressed in a cotton dress with a belt (*chitenje*), they receive the last recommendations and instructions from everyone, without saying a word (often with a handkerchief over their mouth), showing respectful behaviour. The feast is so important that, in some cases, it may be performed even without the *chiputu* and, therefore, for intact girls, as is the case for Christian girls who do not obey this local tradition.

From this moment onward, the girls continue the manipulation privately (Fig. 3), but less and less often until marriage. The function is to confirm that the modification is performed, preventing the structures no longer being manipulated to return to their original morphology.

Consequences for the Woman's Health

From the analysis and comment in Table 1, we see that labial elongation has some positive functions in the lives of Malawi women. In particular, the elongated labia "help the penis to enter the vagina easily and simply," they "help the woman when giving birth," they "provide a large space for the baby to pass through," they "contract the vagina after deliver," and they normally "protect the vagina from dust and germs."

The teenagers, however, also pointed out the negative side. There is the pain of the first manipulation, the additional pain for those who undergo late manipulation,

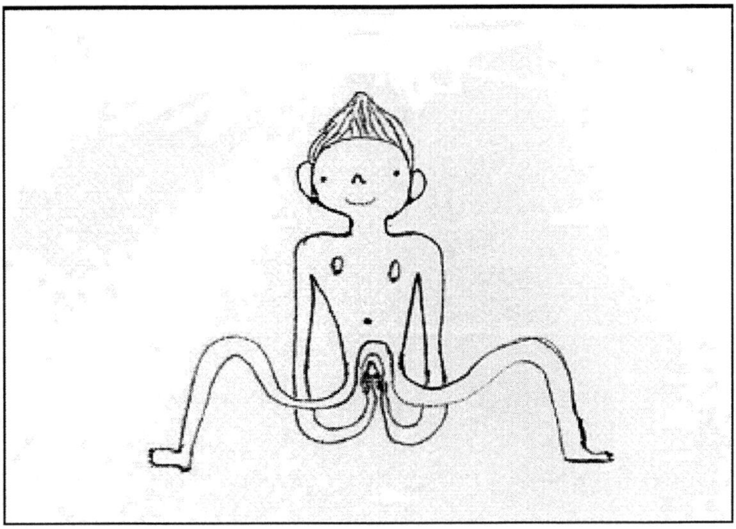

Fig. 3 Automanipulation at home

which stretch structures that are no longer elastic — "Those who stretch their labia before marriage and are older feel pain." "After ritual segregation, girls are encouraged to meet with boys, and this can easily spread AIDS." If the manipulation lasts too long, the labia minora can become too long and therefore a nuisance during everyday life — "One can't walk and sit down, and water comes out of the labia while one is sitting." In addition, labial elongation increases normal vaginal lubrication. The labia "produce water and the woman will always look dirty," "The labia can tie." (They deal with some of these problems by making a knot with the labia themselves, Monica Ottarda, personal communication 2005). Finally, the elongated labia may be a source of discomfort among teenagers — "Girls who haven't stretched their labia minora are laughed at by their friends." "When we are with friends who have stretched, we don't feel comfortable." This discomfort is shown in two of the drawings by two Malawi teachers.

These are the main consequences of labial manipulation on women's health. As seen in Table 1, many of these traits also can be found in Uganda, where, however, there is a peculiar way of stretching the labia minora. A special kind of harness, a belt with leather suspenders, stretches the labia minora with every step. This tool, which can result in torture, seems to be used when the labia minora do not respond adequately to manipulation because the girl did not follow the timetable.[17]

On the whole, however, the consequences of genital stretching on health are not comparable to those of the reductive types of FGMos (excision, infibulation, etc.). This is the reason why international agencies are not interested at all in the first type. Therefore, this type is more or less unknown because scholars do not study how genital stretching influences health. Moreover, the consequences are worse for immigrant women in contact with intact women, as well as Western men, who are not "educated" about how to deal with sexual intercourse with women with exceedingly abundant sexual structures and, therefore, despise them (information from anecdotal stories told by Italian men who had intercourse with these African women).

As an example of the above, we can report the alleged consequences of genital stretching as presented as evidence to support an asylum request by an Ugandan woman in the United States.[2] The Ugandan woman told a gynecologist and a psychologist visiting her that she had problems wearing underwear and pants and being seated was uncomfortable. Ulcers were forming on the manipulated area, which caused discomfort during intercourse, irritation and pain during urination. From a more psychological point of view, she had intrusive and/or disturbing memories connected with the manipulation, discomfort and lack of pleasure during intercourse, feelings of insecurity and fear of not being able to have stable affectionate personal relations due to the manipulation, poor self-esteem, and serious depression was diagnosed after parturition, which was successfully treated.

Another consequence of genital stretching on health is possibly connected to the spread of HIV. Recently, Okuonzy (2005)[18] described a project of AIDS monitoring in south-western Uganda, Rakai district, made by American scholars from Columbia University in collaboration with the National Council for Children, Ugandan Ministry of Gender, Labor and Social Development. According to this study, the disease has been waning for the last ten years, but among teenagers (15–19 years old) the chance of being HIV-positive is six times higher among girls when compared with their male peers.

The above results got a lot of attention among the Baganda women of Buganda (southern Uganda, 2005), who consider labia minora stretching as one of their most genuine traditions. Etyang and Natukunda point openly to the stretching ritual (*okukyalira ensiko*) as an important cause of AIDS transmission and to the teachings, which openly encourage sexual promiscuity, transmitted by the traditional operators, the *ssenga*, during the manipulation performed during the rites of passage. Albeit this charge is not agreed upon by everyone locally because the ritual is considered to be simply cultural, without a change in the girls' sexual behavior, our opinion, based on the Malawian girls' comments (Table 1), certainly agrees with Etyang and Natukunda.

Sexuality: Commented Results

A number of aspects connected with sexuality (Table 2) are evident in the interviews and comments to drawings gathered during the fieldwork in Malawi by the Padua Working Group on the FGMos. In particular, labial stretching is considered in a positive way because "The girls try to be in a position to welcome men of every size without feeling much pain," "It gives a man more pleasure." "It makes sex more enjoyable." "It helps hold the penis and facilitates ejaculation." "The prepared girl adds caressing parts." "The men never run away from their women." Stretching "beautifies the vagina" and the woman "feels lighter and looks nice." "She doesn't find problems when she is married and sweet." The stretching is also an element of

Fig. 4 A girl with stretched labia, well composed and ready for sex (Moro 2004/2005[23])

female satisfaction and, therefore, it satisfies both partners (Fig. 4). When a man undresses a woman, and he sees the stretched labia, he plays with them, he touches them, and he is very excited by the woman."

Some negative aspects for the woman, however, also have been mentioned. "After the preparation, the woman feels different because the physical appearance of the labia differs." "It makes a woman stretch the labia minora each and every day." "… The men and the boys of Yao don't want and don't like stretched labia, they like them natural, without stretching." Female gratification connected with stretching is also mentioned indirectly from the disadvantages of the intact woman, who is marginalized. "Girls who do not have stretched labia minora are laughed at by their friends." "When it comes to sex, sex becomes difficult because there is a small space and so it is very difficult for the penis to enter the vagina." (Fig. 5) Also, men have problems during intercourse, "When a woman or a girl wants to have sex and she hasn't stretched her labia minora, the penis feels pain and has difficulty doing its work because it comes out."

We should note that in Malawi a real sexual culture exists, thus labial stretching is not only a form of socially accepted masturbation, but also a pretext for encouraging sexual promiscuity, mediated through the teachings given to the girl during the coming-of-age rites of passage. In the same context, specific songs are learned and movements of sexual intercourse are rehearsed.

Boys receive parallel teachings, called "techniques of bedwork." They learn about caressing and manipulating the stretched labia and the pubes before introducing the

Fig. 5 A young girl without stretched labia, feels shy and isolated (Moro 2004/2005)

penis, and that, during intercourse, it is the male partner who manipulates the woman. "The man plays with the labia before having sex, it's a kind of masturbation for him." Men like these stretched labia minora because they like to touch them in intimacy with their partners. These labia increase pleasure for both partners and, thus, they favor sexual intercourse, with concurrent increased possibilities of conception (Fig. 6).

Prolific and fertile women are highly appreciated in Africa. The examples in Fig. 6 and Fig. 7, drawn by two Malawian teachers, who represented sexual intercourse with female partners with and without stretched labia minora, respectively, together with their comments, are highly evocative and do not need further comment. Thus, the same psychosexual meaning of the ritual, pointed out for Uganda, is also confirmed in Malawi.[5,6] A similar sexual behavior would be unthinkable for northern and peri-Saharian peoples who presently perform diverse reductive forms of FGM (excision, infibulation, etc.). In Africa, in fact, two opposite sexual cultures occur — the peoples that perform reductive forms of FGM are hostile to the sexual sphere, and psycho-linguistic studies[6] confirm this behavior with scarce or no specific sexual words. In Somalia, a woman must be as stiff as a wooden stick during intercourse and, in any case, she must show no reaction. Female genitals also have a socially negative connotation.

In contrast, in central and southern Africa, where expansive intervention on female genitals is endemic, sexual culture and related terms are still expanding. Thus, even teenagers seem foulmouthed according to Western standards, and their comments as well as their drawings are extremely explicit. In fact, the function of

Fig. 6 A couple making a romance, ready for sex. A woman is playing with man's penis and a man is playing with labia ready for sex (Moro 2004/2005)

Fig. 7 A woman without stretched labia leads to unhappy marriage (Moro 2004/2005)

the ritual is to encourage sexual intercourse and concurrent likely conception. Paradoxically, both reductive and expansive societies believe that the modifications favor women's fertility, and the modifications themselves are considered to be an indispensable premise to marriage (Fig. 7).

Conclusion

The analysis on the Malawian data shows that FGMo consists of the ritual and symbolic re-modeling of the whole perineal region, both indirectly and directly connected with fecundity. It is accompanied by physical and cultural maturity, which makes the manipulated young woman always adequate to all aspects of her culture.

The comparison with the meanings this same tradition carries in Uganda[5,6] shows overlapping results, with only one difference. In fact, the Baganda did not make explicit the negative traits that emerged in Malawi, where the encounter and mixture of different ethnic groups gives specific results, especially psychological aspects. The sharing of the same environment by intact and manipulated girls who often live in the same village appears to determine envy for the former and discomfort for the latter.

There is no doubt that FGMo, as it is performed in Malawi, is an example of the fourth type of Female Genital Mutilation,[15] yet, this country does not appear on any list, not even recent, that describes the phenomenon as endemic.[19] As far as we know there are no fieldwork data (not even anecdotic) on the prevalence of labial stretching. We believe that, even today, the expansive forms of FGMo in Africa, rarely studied in depth in the past, have not stirred any scholarly interest yet, either locally or internationally. Nevertheless, these manipulations do not appear light, and are certainly more and more important if we consider them in the African women's diaspora.[17] We hope, therefore, that our study may increase the interest in carrying out scholarly work on this subject in Malawi as well as in other African countries.

References

1. Nervi L (1994) Malawi: fiamme nel cielo africano. Velar, Bergamo.
2. Grassivaro Gallo P, Tita E, Viviani F (2006) *At the Roots of Female Genital Modifications (FGMo)*. In Bodzsar EB & Susanne C. (Eds) Human Evolution: Facts and Factors. Biennial Books of European Anthropology Association (EAA), Eotvos University Press, Budapest, 4:159–169.
3. Moro D, Grassivaro Gallo P, Passaquieti A (2006) Female Genital Modifications in Malawi. A Field Research. Paper presented at X International Symposium on Circumcision, Genital Integrity and Human Rights, National Organization of Circumcision Information Resource Centers (NOCIRC) & Doctors Opposing Circumcision, University of Washington, Seattle, 22–24 August.
4. Lis A, Mazeschi C (2005) Emozioni e rappresentazioni nel disegno del bambino. Aspetti teorici relativi al significato del disegno quale espressione del mondo interno del bambino. Paper presented at Programma di Formazione 2005, Azienda Ospedaliera Sant'Anna, Como, 20 May.
5. Grassivaro Gallo P, Villa E (2004) Longininfismo rituale tra le Baganda (Uganda). Rapporto Preliminare. Rivista di Sessuologia, 28:17–22.
6. Grassivaro Gallo P, Villa E. *Ritual Labia Minora Elongation Among the Baganda Women of Uganda*, "Psychopathologie Africaine," 33:213–236, 2005/2006.
7. Chakanza JC (1995) The Unfinished Agenda: Puberty Rites and the Response of the Roman Catholic Church in Southern Malawi, 1901–1994. In: Anonimous Authors (eds.) Religion in Malawi. Department of Theology and Religion Studies, n. 5, Chancellor College, University of Malawi.
8. Janssen DF (2004) Growing up Sexually. Vol. 1, World Reference Atlas. 0.2 Magnus Hirschfeld Archive for Sexology, Berlin.
9. Sardella PM (June, 2003) Lomwe Anthropology: Religious Tradition and Culture of the Alomwe of Phalombe District, in Malawi. Thesis for Licentiate, Pontificial Gregorian University, Fac. of Missiology, Rome.

10. Franceschini L (ante 1998, s.d., anno di morte dell'A.). Appunti sul Malawi. Pro manuscripto.
11. Ott M (2000) African Theology in Images. Kachere Monograph, n. 12, Christian Literature Association in Malawi (CLAIM), Blantyre, Malawi.
12. Phiri IA (1997) Women, Presbyterianism and Patriarchy: Religious Experience of Chewa Women in Central Malawi, Christian Literature Association in Malawi (CLAIM), Kachere Monographs, n. 4, Blantyre, Malawi.
13. Van Breugel JWM (2001) Chewa Traditional Religion, Christian Literature Association in Malawi (CLAIM), Kachere Monograph, n. 13, Blantyre, Malawi.
14. Kunt KE. Vaginal Drying Agents and HIV Transmission, International Family Planning Perspectives, 24, 2. In: www.agi-usa.org, January 6, 1998.
15. World Health Organization (1996) FGM: Report of the WHO Technical Working Group, July 17th–19th 1995, Geneva.
16. Boeder RB (1984) Silent Majority. A History of the Lomwe in Malawi, African Institutd of South Africa, Cape Town.
17. Weinberg L, Mayer Fuchs C, Ccondon J-B, Leigh Hobbs J, Ortiz A (2004) Supplemental Pretrial Brief in Support of Respondents' Application for Asylum, Withholding of Removal, and Withholding of Removal Under the Convention against Torture. U.S. Department of Justice Executive Office of Immigration Review. Immigration Court Boston (MA).
18. Okuonzy A (2005) Pragmatic safe sex, not Abstinence or Faithfulness, was key in Uganda's HIV Decline. Health Policy and Development, 3. April, 1 (ital translation: Donà S. La riduzione dell'HIV/AIDS in Uganda. Salute e Sviluppo, 3:7–9, 2004).
19. Yoder SP (2004) Challenger in Collecting Data on FGC. Paper presented at the Seminario Mutilazioni dei Genitali Femminili. Una Giornata di Riflessione, AIDOS, June, 22, Rome.
20. Etyang J, Natukunda C (2005) "Visiting the bush" LINKED TO hiv. The New Vision of Uganda's leading daily. Kampala (posted to the web, August 16).
21. Grassivaro Gallo P, Villa E (2006) Psycholinguistic Approaches to *Ritual Labia Minora Elongation among the Baganda Women of Uganda*. Chapter 5, pp. 57–64. In: 147- Denniston GC, Grassivaro Gallo P, Hodges FM, Milos, MF, Viviani F. Bodily Integrity and the Politics of Circumcision: Culture, Controversy, and Change Springer (New York,USA); 2006.
22. Phiri M (1993) Girls Initiation Rites (Utchato) among the Amihavan (Lomwe), People. Paper Research, Queen of Apostles Inter-Congregational Seminary (I.C.S.), Mangochi, Malawi.
23. Moro D. Modificazioni Genitali Femminili in Malawi. Resoconto di un'indagine sul campo. Tesi di Laurea in Psicologia. Università di Padova. AA 2004/5.

Variations in Penile Anatomy and Their Contribution to Medical Mischief

Ken McGrath

Abstract It is clear that the human penis is as individual as a fingerprint, so that no two are alike. Some of the variations have been used as justifications for circumcision, but are they common or genuinely pathological? And can the modern standardized methods ("one size fits all") provide a predictable result in the face of such variation or any therapeutic value? Three of these variations will be reviewed for their nature and significance.

Introduction

In spite of a number of recent studies, ignorance of the skin system of the human penis remains widespread both within the medical professions and the public. Sadly, there is not much help to be found in the standard texts or medical dictionaries, most of which continue to reprint the falsehoods of the past. We should not be surprised, therefore, to find parents and physicians frequently worrying about the state of the penis in young males. This concern often gives rise to either an intention of inappropriate correction of a normal structure or obsessive meddling with a healthy organ. The development of the human penis is a complex sequence of events that results in an utterly individual outcome: no two penises are identical and there is a surprising range of anatomical detail that should be considered normal. Parents need to be assured of this range of normality. Moreover, the desire of practitioners for standard procedures can lead to unpredictable outcomes because of both this anatomical variation and the impossibility of predicting the functional results of surgical correction in infants. This is because the procedures are performed with an emphasis on achieving an acceptable cosmetic outcome on a very small organ that has the capacity for considerable growth and changes during puberty. Furthermore, no surgical procedure can have absolutely predictable outcomes because of the variations in healing and scar formation, the individual variations in technique, and the effects of infection.

Senior Lecturer in Pathology in the Faculty of Health, Auckland University of Technology, New Zealand

G. Denniston et al. (eds.), *Circumcision and Human Rights*,

Regrettably, it seems that the majority of those performing surgical procedures on the penis of minors take no interest in following up the outcome after the organ has developed. Admittedly this involves a time span of at least a decade, but there have been very few attempts at quality control and many boys are left with seriously damaged penises; the outcome is often a functional impairment and is seldom tidy to say the least. Three anatomical variations are presented here as examples of unnecessary concern that have engaged the medical mind since 1694. Two are sources of, and the continuing motivation for, the most serious medical mischief — circumcision.

Sebaceous Glands

The existence of these glands in and around the inner prepuce and coronal sulcus has been reported for over 300 years — since William Cowper inflated into a fact[1] a scrawled note made by Edward Tyson in his handwritten lecture notes on the glands of the abdomen.[2] Under the heading "Urethra Ejus," Tyson inserted "Glandula Mucilaginosa," together with a footnote stating that he had lately discovered these. We have no idea to which animal he was referring, the details of these glands, or to whom the note was addressed. Ironically, given the context, it may have been that Tyson discovered the glands now attributed to Cowper! On the other hand, it is also possible that Tyson saw the glands imbedded in the distal urethra reported by Littre in 1703.[3] Within a few years, Cowper published a passage[1] that stated Tyson had found small glands where the prepuce joined the glans, which gave off an acrid and odiferous liquor capable of corroding the glans if retained by a long foreskin. Thus were born *Tyson's Glands,* the existence of which has become one of the longest held myths of medicine, described in every anatomy textbook since 1694.[4] These glands were quickly identified as the source of smegma in the early eighteenth century and, therefore, a source of danger in light of Cowper's description in individuals possessing a long foreskin. With the rise of the hygiene mania in the nineteenth century, they and smegma became a primary justification for widespread circumcision. The debunking of the existence of these glands in the human penis by Keith and Shillitoe[4] in 1904 has been confirmed by one paper that set out to elucidate what Tyson might have seen (the findings are relevant to the next topic below),[5] and another that searched for them whilst analyzing the nature of smegma.[6] Here I must admit to an error in a paper published in 2001 in which I mentioned this myth,[7] but wrongly cited a publication of Tyson's on the Orang-Utan[8] (actually an infant chimpanzee whose skeleton still exists) as Cowper's source. Many of the modern accounts of *Tyson's Glands* also make this erroneous association. Having again considered this long-standing anatomical myth, I will pass on to another revision. In our 1999 paper,[9] Christopher Cold and I reported that we could find no evidence for these glands in the human prepuce or the coronal sulcus; this confirmed the findings above. Thus, for some time I have been content with the belief that no glands existed in the sub-preputial space and, by extension, in the skin of the prepuce as a whole. Supporting this is the fact that the distal skin of the penis (which includes the prepuce) is glabrous (hairless) and that sebaceous

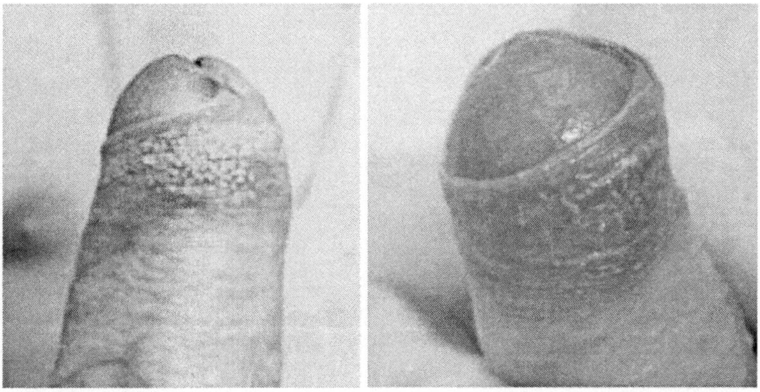

Fig. 1 'Ectopic' sebaceopus glands in the preputial transition zone of a young European male

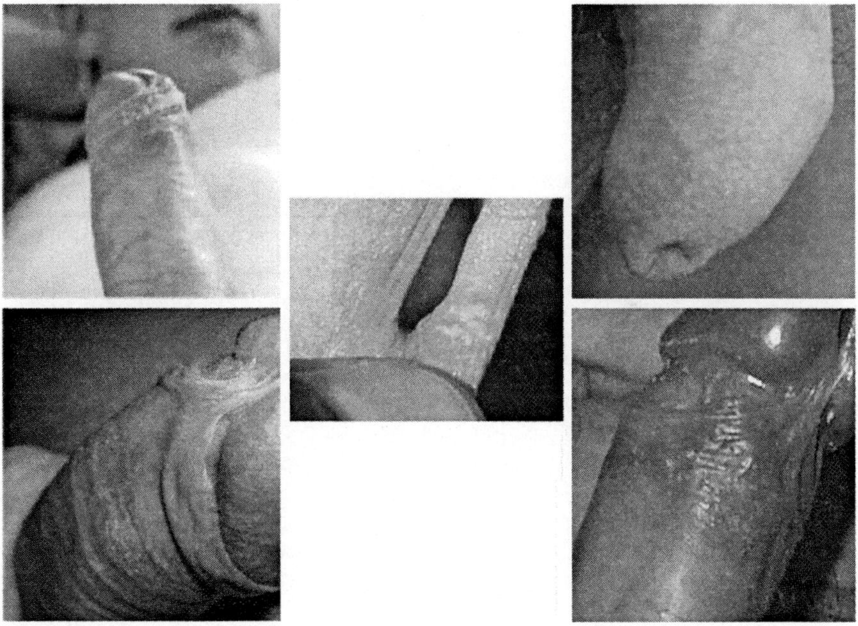

Fig. 2 Further examples of 'ectopic' sebaceous glands in the preputial transition zone

glands are usually associated with hair follicles where they lubricate the emerging hair shaft and oil the surrounding hirsute skin.

A year ago, I noticed a group of obvious ectopic sebaceous glands distributed across the preputial transition zone of a young European male (Fig. 1). Having been alerted to this possibility, more were subsequently identified in a few other individuals (Fig. 2). Co-incidentally the existence of the more common ectopic sebaceous glands in the margins of the lips (another transitional zone) in about 80–95% of adults was found in reports that noted they are named *Fordyce Spots*.[10–12] Included in a description of these benign spots, which entirely matched those I had seen in the prepuce, was a note that they have been found in other mucosal transitional zones in both genders. A search of the literature discloses that these ectopic sebaceous glands or Fordyce's spots on the penis have been previously reported.[5,13–15] The existence of "displaced" and hairless sebaceous glands (or milia) in the distal glabrous skin of the penile shaft seems to be well known and probably is not infrequent, especially in the ventral skin where they may be a hairless extension of the Area of Klatsch.[7] Because the skin of the penis is thin, these glands are easily seen as small, yellowish, sub-epithelial, dome-shaped papules about the size of millet with a shiny surface when the skin is stretched. However, they seem to be uncommon in the preputial junction zone (probably about 1:500–1,000) and they do not appear to be found in the sulcus or around the frenulum, which discounts any possibility of them being the elusive *Tyson's Glands*. Only two reports of them being discovered on the glans were available in the literature,[15,16] so we must suppose that this is very rare. It does not appear that circumcision eliminates ectopic sebaceous glands.[17]

An analysis of the chemical state of smegma[18] failed to find any evidence for sebaceous secretion products, which suggests that the Fordyce spots do not contribute to any sub-preputial accumulation. A number of authors[5,6] have supported the opinion of Keith and Shillitoe[4] that smegma is not a secretion rather a product of transudate and desquamated cells. Smegma has a long history of bad reputation thanks to Cowper's comments, and all of the suggested dangers were progressively discounted in the twentieth century. A very recent review continues this progress by effectively disposing of the supposed carcinogenicity of smegma.[19] Another myth, so often encountered, is the statement that collection of smegma under the foreskin is inevitable, especially when washing is neglected. A number of authors have commented on its relative rarity and one study of 1,000 males[6] made a quantitative evaluation of its presence, finding that it was absent in 60%, and thinly visible in 20%, of men without phimosis. The surprising finding was the absence of smegma in 86% of men with phimosis, which was quite the opposite of the generally held opinion. None of the individuals depicted in Figs. 1 and 2 had any evidence of smegma.

Pearly Penile Papules

The second curiosity to consider is the rather odd collection of white nodules not infrequently observed distributed around the corona glandis or in the sulcus on both sides of the frenulum (Fig. 3). The latter site may well have given credence to the existence of *Tyson's Glands* over the centuries, but they are not associated with any glandular structure. These nodules certainly look anomalous and have been the cause of concern to young males from time immemorial. It is, therefore, strange that, in spite of their entering the medical literature in a report by Littre and Morgagni in 1700,[20] their histological nature was resolved 300 years later.[5,21,22] Since that time, they have appeared from time-to-time in reports making simple observations in man and speculations such as their being a nerve organ.[23] The prominent neuroanatomist Bielschowsky was able to show in 1909 that no nerve structures were associated with them in man and this has been confirmed.[22]

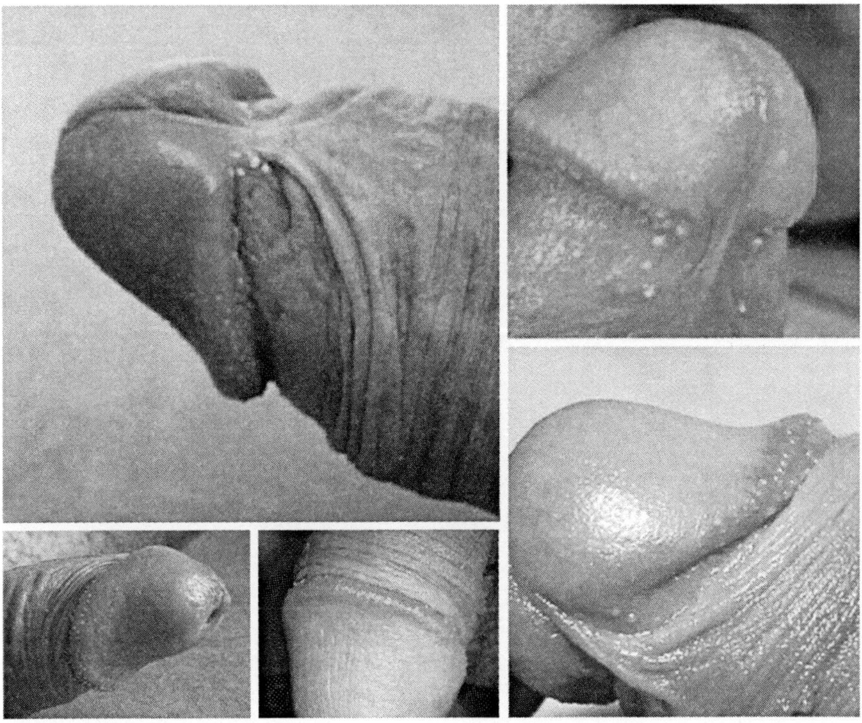

Fig. 3 Examples of Pearly Penile Papules, showing the range of incidence from a few near the frenulum, through a few scattered across the corona, to involvement of the whole corona

Buschke[24] seems to have been the first anatomist to have compared their infrequent occurrence in man with their normal appearance in other mammals, such as felines, pigs and primates. We have shown that, in the chimpanzee, these papules are a normal feature (spine-like) and are associated with nervous structures.[9] It seemed to us that, in man, they may be a return to an earlier morphology. A paper of the same year also suggested that these human papules may be a phylogenetic residuum.[25] As recent as 1996, they were still being described as lesions, which is odd, to say the least, given that so many recent studies have made a definite determination of their non-pathological nature, some concluding that they were better regarded as harmatomas. The name *pearly penile papules* seems to have been coined in 1964,[21] and has found more favor than the terms *hirsutoid papillomas*, *papillomatosis coro nae glandis*, and *papillomatosis penis*, used in a minority of papers since 1969. It is clear that physicians (and patients[25]) have also worried about the nature of these papules for some time, some of them making observations of their incidence and going as far as attempting to treat them like warts, using podophyllin (to which there was no response),[5] electrodesiccation or curettage (with a high risk of scarring),[26] and liquid nitrogen or CO_2 laser (which had satisfactory results).[25,26] They have been shown to have no evidence of glandular structures[5] or of HPV infection,[27] and the consensus of recent papers is that they should not be treated as they have no evidence of pathology.

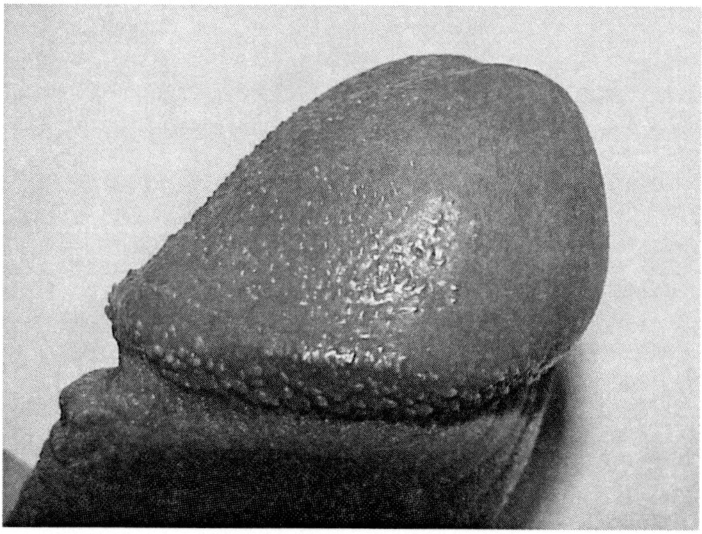

Fig. 4 A very rare extension of PPP onto the dorsal glans

It is clear from even a cursory set of observations that the number or extent and position of these papules varies considerably. Some males have only two to four and others have so many that the entire corona is covered with them (Fig. 3). One case (Fig. 4) is the most extensive I have seen, and it can be seen that the papules not only cover the corona, but also extend over much of the dorsal surface of this man's glans; this must be rare, as I have not seen any others like it nor read any other account of such a spread. While Buschke reported an incidence of 6% prominent and 8% total papules in 435 subjects in 1909,[24] later papers have found much higher incidences; e.g., 14% in Singapore,[26] 20% of 229 subjects in Texas,[22] 30% in 840 clinic patients in Detroit,[27] 33% in 642, 19-year-old Swiss men,[28] and 48% in 200 clinic patients in Cambridge.[25] The last series in Cambridge was also counted by the extent of coverage of the corona: few and small, 35%; small and extending around the corona, 9%; moderate to large and extending around the corona 4%.[25] There is debate about the age of appearance, with some authors stating an onset just prior to puberty, but they seem to be rare in early childhood, having escaped the notice of pediatricians and having been reported only once in a neonate.[29]

Redundant Foreskin

The third curiosity to consider is the oft-diagnosed redundancy of the prepuce. In most of the developed world, removal of tissue in a hospital requires the recording of a diagnostic code for statistical analysis, patient records, and financial accounting. By far, the most common code used to justify circumcision is *ICD 605: redundant prepuce and phimosis*. There is no need to discuss here the erroneous diagnosis of phimosis in the majority of healthy infants, which has been thoroughly reviewed, but the description "redundant" certainly bears examination. Dictionaries usually give two definitions of this adjective: (1) abundant, copious, excessive; and (2) spare, surplus, turning to advantage. I would suggest that the justification used by surgeons is from the first; i.e., "excessive," or hypertrophic, and that is borne out by the writings of the nineteenth and early twentieth centuries.[30,31] We could counter with the questions: (1) how is this judged in respect of what may be considered normal? and (2) how can it be judged in light of the growth changes that will occur in the penis with puberty? The diagnosis is essentially a presumption; as George Bernard Shaw wrote: "When an operation is once performed, nobody can ever prove it was unnecessary." All boys have an *acroposthеon* at birth,[32] except in those with a hypospadias, and the second definition of redundant as "turning to advantage" is surely the correct one in this context, as it provides extra skin for later growth.

The most interesting question on preputial redundancy, or the acroposthеon, is to ask if it is retained into adulthood. Although the literature is silent on this, simple observation discloses that a significant number of men retain their acroposthеon (Fig. 5).

While it is impossible to suggest an incidence for this at present, the reasons for retention can be explored. One of the first to notice types of penile shape was Dickinson, who produced an unique atlas of genital anatomy from his lifetime of studies.[29] One of his drawings illustrates the fact that the shape of the human glans takes two forms when seen laterally: a square or blunt type, whose coronal axis is

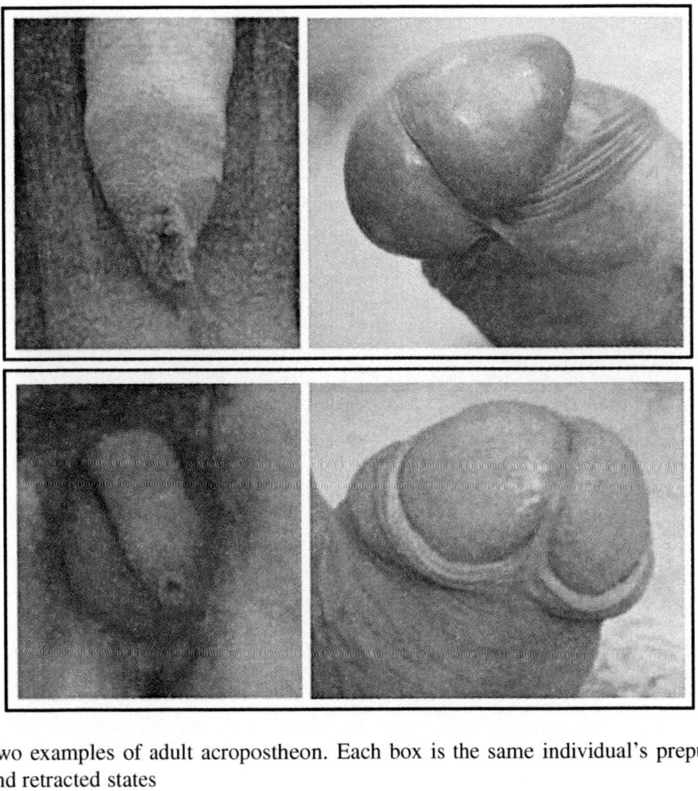

Fig. 5 Two examples of adult acroposthеon. Each box is the same individual's prepuce in the relaxed and retracted states

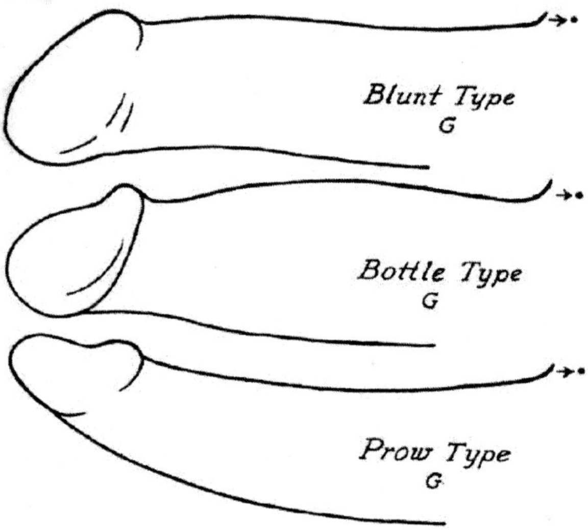

Fig. 6 Penile classifications from Dickinson's Atlas of Human Sex Anatomy (2nd edition, 1949)

more or less perpendicular to the shaft, and a prow shaped or more pointed type, whose coronal axis is oblique to the shaft to various degrees (Fig. 6). My own observations of a large cohort suggest that this difference is one of the two determinates of acropostheal retention. The other determinate of retention is the state of the frenulum, which should come as no surprise as it is generally accepted that, although the frenulum forms from completion of the penile raphe, it does have the commonly observed functions of tethering and replacement of the prepuce. The frenulum varies significantly between individuals, taking a number of states:

1. Absent in a normal penis or unformed in mild hypospadias (Fig. 7).
2. Very short in *frenulum breve,* which prevents complete retraction of the prepuce and generates chordee of the glans with traction at the full extent of available retraction (Fig. 8).
3. It terminates in the sulcus or at the very edge of the corona (Fig. 9).
4. It terminates well up the glanular grove, at or close to the meatal margin (Fig. 10).

Frenulum breve is a well-recognized condition in urology, as are the various degrees of hypospadias, but there are no terms available to describe states 3 and 4. I, therefore, propose that they be known as *low frenulum* and *high frenulum,* respectively. It seems clear from our observations that, for the most part, men who develop a blunt glans with low frenulum have a shorter foreskin and no acropostheon, and those with a prow glans and high frenulum have a long foreskin and acropostheon. Indeed, the greater the angle of the corona to the shaft (and therefore the dorsal length of the glans) and the nearer the attachment of the frenulum is to the meatal margin, the more extensive the acropostheon (and the length of foreskin). Furthermore, those individuals in this state of high frenulum and acropostheon will also have a more

Fig. 7 Examples of incomplete development of the frenulum (centre), and mild hypospadias (left and right)

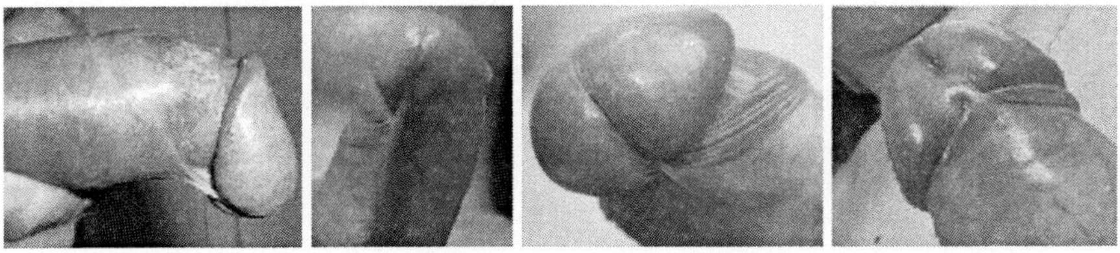

Fig. 8 Examples of frenulum breve, also showing chordee (far left and centre left)

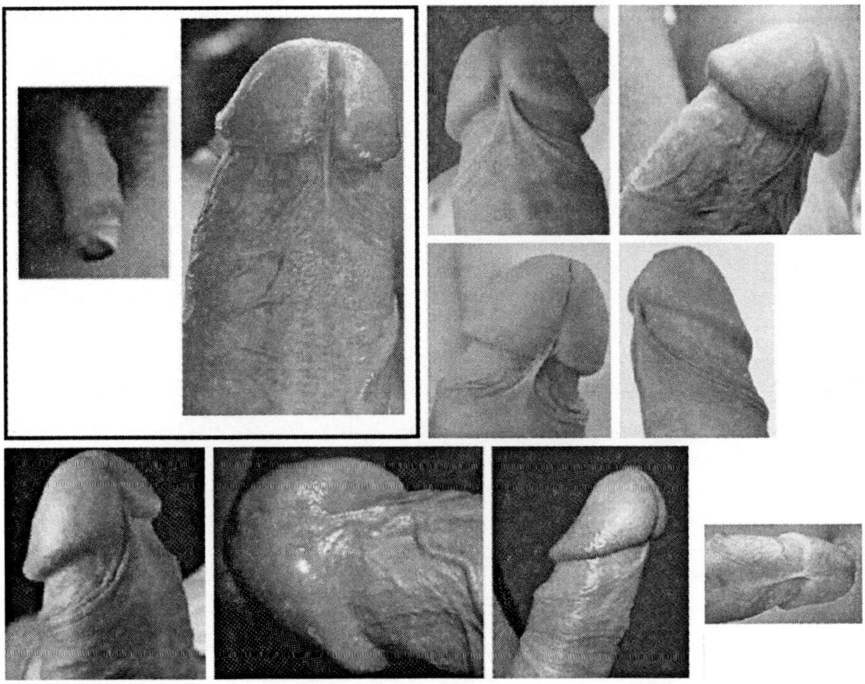

Fig. 9 Examples of the frenular appearance in individuals with no acropostheon. The 'Low' frenulum with a blunt glans in most cases

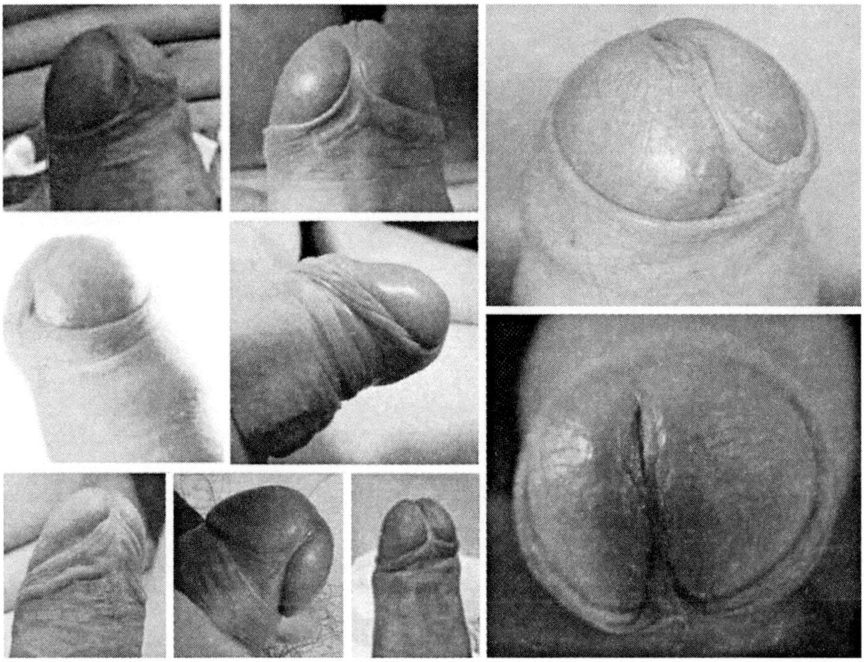

Fig. 10 Examples of the frenular appearance in individuals with an acropostheon. The 'High' frenulum with a prow glans

extensive frenular delta.[7] The age at which these determinants become settled during development is unknown, so it is presently impossible to say if they can be predicted in infancy.

Taking these anatomical differences into consideration, it will be apparent that the standardized methods of circumcision using bell devices will give very different outcomes with respect to the amount of inner prepuce that is preserved and the position of the resultant scar; this is because of the physical factors of the angle of the corona and the extent of the frenulum. This may be the reasoning behind the practice of some surgeons who ablate the frenulum as part of their circumcision procedure. Moreover, those boys who would have retained their acropostheons are more likely to have more inner prepuce survive the circumcision procedure if their frenula are not destroyed. Each operator applies the procedure idiosyncratically and without any absolute parameters, to which must be added the fact that it is surgery on a very small organ that will undergo a large increase in size (usually at least six times) at puberty. These factors may account for the wide variation in circumcision outcomes. A major failure in quality assurance with circumcision is that practically all operators never make any assessments of the outcome of their surgery after the period of growth so as to reduce the oft-seen poor results. It is left to other practitioners to repair the disasters and problems as they present later.

Summary

The following statements can be made:

1. "Tyson's Glands" do not exist in the human penis.
2. Ectopic sebaceous glands or Fordyce Spots are occasionally seen in the ventral shaft skin and rarely in mucocutaneous junction of the prepuce of the human penis. They do not contribute to smegma, which is not a secretory product.
3. Pearly papules are commonly (14–48%) found projecting from the corona of the human glans and/or in the sulcus adjacent to the frenulum. They are not pathological, but a harmless angiofibroma or harmatoma and probably a phylogenetic residuum.
4. Preputial redundancy in childhood does not exist because the "extra" skin has a purpose in later developmental growth.
5. Because of the number of complex variables in fetal development, each penis is as individual as a fingerprint, i.e., no two are exactly alike.
6. Retention of an acropostheon in adulthood is dependent on both the shape of the glans and the length and/or the point of attachment of the frenulum.
7. "Standardized" circumcision cannot give a predictable result and the procedure is notorious for its lack of assessment of outcome by its practitioners.

Acknowledgments The author is grateful to Dr. Robert Darby for help with sources and scholarly support, and to Martin Novoa for extensive discussion on the topics of this paper.

References

1. Cowper W. Myotomia Reformata: Or, A New Administration of all the Muscles of Human Bodies. London: Sam. Smith and Benj. Alford, 1694. pp. 228–9.
2. Montagu A. Edward Tyson MD and the rise of human and comparative anatomy in England. Mem Am Philos Soc 1943;20:206.
3. Littre A. Description de l'urèthre de l'homme. Histoire de l'Académie royale des sciences avec les Mémoires de mathematique et de physique. Paris: n.p. (1700), 1719. pp. 311–6.
4. Keith A, Shillitoe A. The preputial or odoriferous glands of man. Lancet 1904;1:146–8.
5. Hyman AB, Brownstein MH. Tyson's "glands": ectopic sebaceous glands and papillomatosis penis. Arch Dermatol 1969;99:31–6.
6. Parkash S, Jeyakumar S, Subramanyan K, Chaudhuri S. Human subpreputial collection: its nature and formation. J Urol 1973;110:211–2.
7. McGrath KA. The frenular delta: a new preputial structure. In: Denniston GC, Hodges FM, Milos MF (Eds). Understanding Circumcision: A Multi-Disciplinary Approach to a Multi-Dimensional Problem. New York: Kluwer/Plenum, 2001. pp. 199–206.
8. Tyson E. Orang-outang, sive, Homo Sylvestris: or The Anatomy of the Pygmie compared with that of a Monkey, an Ape and a Man: with an Essay Concerning the Pygmies, the Cynocephali, the Satyrs and Sphinges of the Ancients. London: Thomas Bennet and Daniel Brown, 1699 (a review of this book and its 2nd edition may be viewed at: http//:www.kcl.ac.uk/depsta/iss/library/speccoll/bomarch/bomnovdec.html).
9. Cold CJ, McGrath KA. Anatomy and histology of the penile and clitoral prepuce in primates: evolutionary perspective of specialised sensory tissue of the external genitalia. In: Denniston GC, Hodges FM, Milos MF (Eds). Male and Female Circumcision: Medical, Legal, and Ethical Considerations in Pediatric Practice. New York: Kluwer/Plenum, 1999. pp. 19–29.

10. Fordyce JA. A peculiar affection of the mucous membrane of the lips and oral cavity. J Cutan Dis 1896;14:413–9.
11. Monteil RA. Les grains de Fordyce: maladie, heterotopie ou adenome? Etude histologique et ultrastructurale. J Biol Buccale 1981;9:109–28.
12. Dreher A, Grevers G. Fordyce-Flecken. Ein wenig beachteter befund im bereich des lippenrotes und der mundschleimhaut. Laryngo-Rhino-Otologie 1995;74:390–2.
13. Goeckerman WH. Fordyce's condition of the penis. Arch Derm Syph 1926;14:50.
14. Piccinno R, Carrel CF, Menni S, et al. Preputial ectopic sebaceous glands mimicking molluscum contagiosum. Arch Dermatol Venerol 1990;70:344.
15. Saalfeld E. Ueber die Tyson'schen drüsen. Arch Mikr Anat 1898;53:212–8.
16. Massmanian A, Valls GS, Sempere FJV. Fordyce spots on the glans penis. Br J Dermatol 1995;133:498–500.
17. Carson HJ, Massa M, Reddy V. Sebaceous gland hyperplasia of the penis. J Urol 1996;156: 1441.
18. Prakash S, Rao R, Venkatesan K, Ramakrishnan S. Sub-preputial wetness: it [sic.] nature. Ann Natl Med Sci (India) 1982;18:109–12.
19. Van Howe RS, Hodges FM. The carcinogenicity of smegma: debunking a myth. J Eur Acad Dermatol Venerol 2006;20:1046–54.
20. Littre A, Morgagni GB. French Academy of Sciences report. Padua: Adversaria Anatomica, 1700. p. 307.
21. Johnson BL, Baxter DL. Pearly penile papules. Arch Dermatol 1964;90:166–7.
22. Glicksman JM, Freeman RG. Pearly penile papules: a statistical study of incidence. Arch Dermatol 1966;93:56–9.
23. Buschke A, Gumpert H. Die papillen an der corona glandis in vergleichendanatomischer und ethnologischer beziehung. Arch f Frauenk 1925;11:43–55.
24. Buschke A. Über die bedentung der "papillen" der corona glandis. Klin Med 1909;5: 1621–3.
25. Sonnex C, Dockerty WG. Pearly penile papules: a common cause of concern. Int J STD AIDS 1999;10:726–7.
26. Agrawal SK, Bhattacharya SN, Singh N. Pearly penile papules: a review. Int J Dermatol 2004;43:199–201.
27. Hogewoning CJA, Bleeker MCG, van den Brule AJC, Voorhorst FJ, van Andel RE, Risse EK, Starink TM, Meijer CJLM. Pearly penile papules: still no reason for uneasiness. J Am Acad Dermatol 2003;49:50–4.
28. Rufli T, Eichenberger P, Heer K. Papillomatosis coronae glandis. Haufigkeit und klinisches bild. Schweiz Med Wochenschr 1978;108:229–31.
29. Dickinson RL. Human Sex Anatomy: A Topographical Hand Atlas. 2nd edition. Baltimore: Williams & Wilkins, 1949.
30. Hodges FM. The history of phimosis from antiquity to the present. In: Denniston GC, Hodges FM, Milos MF (Eds). Male and Female Circumcision: Medical, Legal, and Ethical Considerations in Pediatric Practice. New York: Kluwer/Plenum, 1999. pp. 37–62.
31. Darby R. A Surgical Temptation: The Demonization of the Foreskin & the Rise of Circumcision in Britain. Chicago: University of Chicago Press, 2005.
32. Bland-Sutton J. Circumcision as a rite and as a surgical operation. Br Med J 1907;1:1408–12. Cited inter alii in Ref 31, p. 228.

The Normal, Natural Penis and the Effects of Circumcision

James L. Snyder

Abstract This paper illustrates many of the unintended injuries caused by circumcision in both in infancy and later. The existence and incidence of these injuries are not widely known, either by the general public or, surprisingly, by physicians themselves.

Introduction

According to the findings of the American Academy of Pediatrics' 1999 task force statement on circumcision, the scientific data is not sufficient to recommend routine neonatal circumcision.[1] Still, the practice continues. Circumcision is practiced in some societies in the world, for various social, religious, or cultural reasons. Usually, it is performed at the request or with the compliance of parents who love their children and who wish to do what they believe is right. Some of the reasons given are that the circumcision will make their children more attractive to a mate, or because it is cleaner, or to prevent some ill condition. These excuses are among the reasons given, whether the child is a male or a female. But, circumcision is usually performed without consideration of the value or function of the natural structures, and with little consideration of the risks of death or deformity to the normal children who are subjected to this procedure. This article is intended to explore these issues and to argue that removal of a child's normal, natural, healthy, functioning sexual structures is not beneficial. While it should be recognized that removal or alteration of the body parts of children, male or female is a violation of their ability to exercise autonomy or free choice for themselves, this article is mostly oriented to the effects of male circumcision for non-religious, non-medical purposes as it is practiced in the United States.

Clifton Forge, Virginia, USA, Diplomate of American Board of Urology, a Fellow, American College of Surgeons, Past President, Virginia Urological Society, Retired Commander, Medical Corps, United States Naval Reserve, retired.

G. Denniston et al. (eds.), *Circumcision and Human Rights*,

The Normal, Natural Human Penis

A diagrammatic view of the penis shows the naturally doubled distal penile skin. The outer layer is continuous with and is structurally the same as the skin of the shaft of the penis. The inner layer is specialized skin not found in any other part of the body.

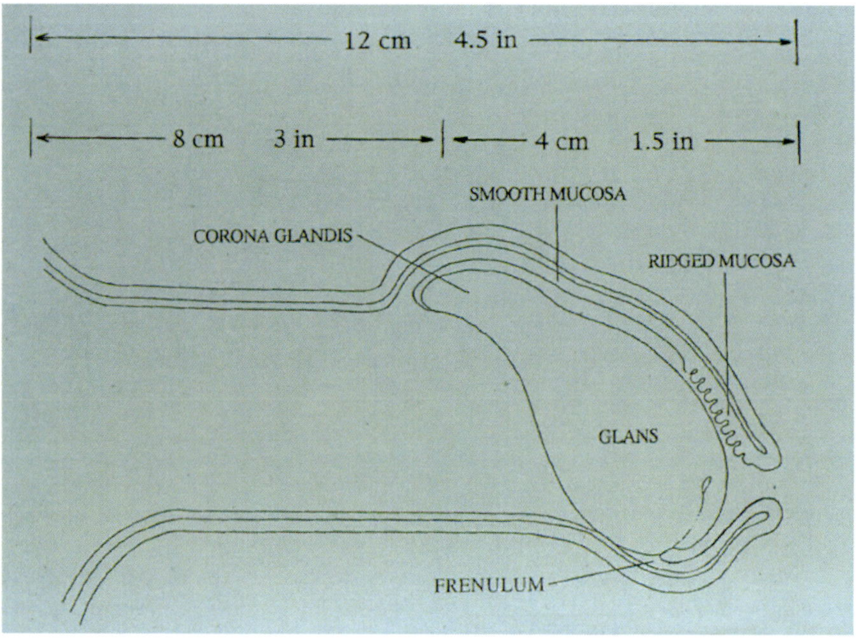

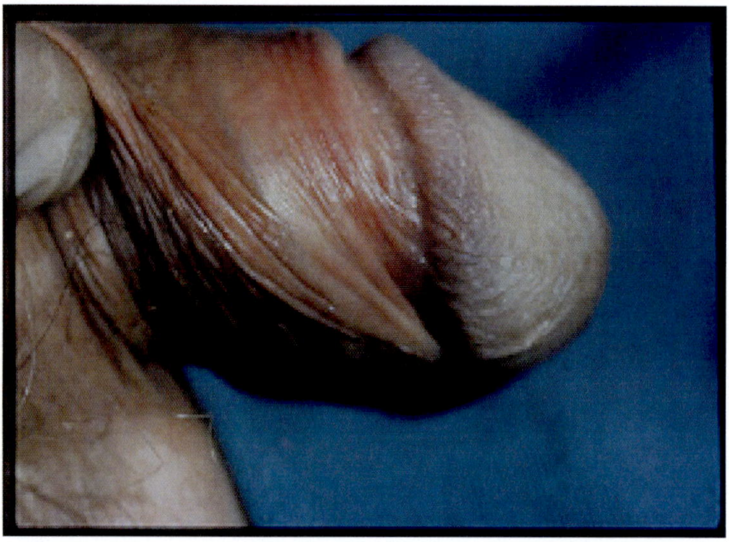

Photo: A normal penis with foreskin retracted. Notice the ridged band of skin with many folds extending from the frenulum and circling the penile shaft. This unique skin is thin, with specialized structure including sensory nerve endings, which enhance sexual pleasure. Unfortunately, circumcision destroys most of this skin with loss of sexual sensation and reduced sexual pleasure.

The normal penis naturally has sufficient skin that moves forward to cover the glans penis. And, the skin is mobile enough that it can be retracted to expose the glans for sexual pleasure, or cleaning, or for any desired purpose. It should be noted that the motion described is in itself pleasurable and is part of the sexual pleasure that is the birthright of any man. This freedom of motion, so that a point of skin can be made to move half the length of the underlying organ is not found in the fingers or toes, and is useful in many of the functions of the intact penis. Understanding these functions will lead to awareness that the foreskin is really not a separate structure which can be removed without effect, but it is essential to the penis in performance of its most important function, the provision of daily pleasure leading to sexual activity and reproduction.

The following series of photographs illustrates the large size of the foreskin and its range of motion along the shaft of the penis. Note the movement of the circular pen marks at the tip of the foreskin and the area of the foreskin over the corona glandis.

These photos show the extent of motion possible when the skin is put on tension during the phases of erection. They also demonstrate that at rest, the distal skin of the penis is doubled under, allowing the most sensitive part of the penile skin and

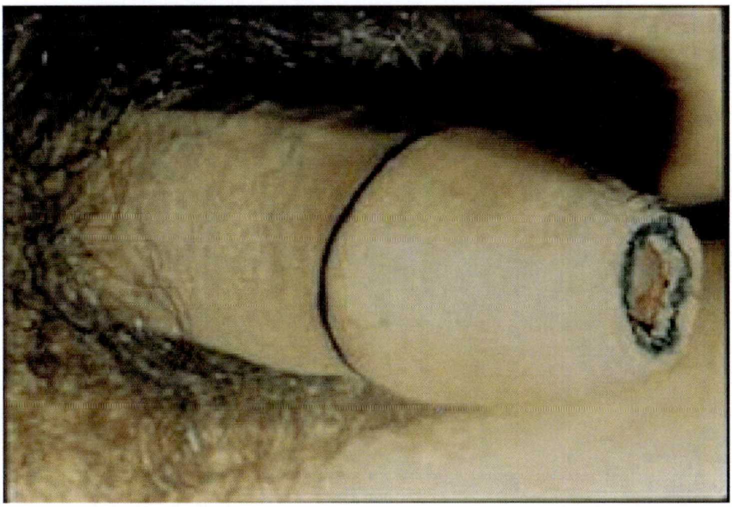

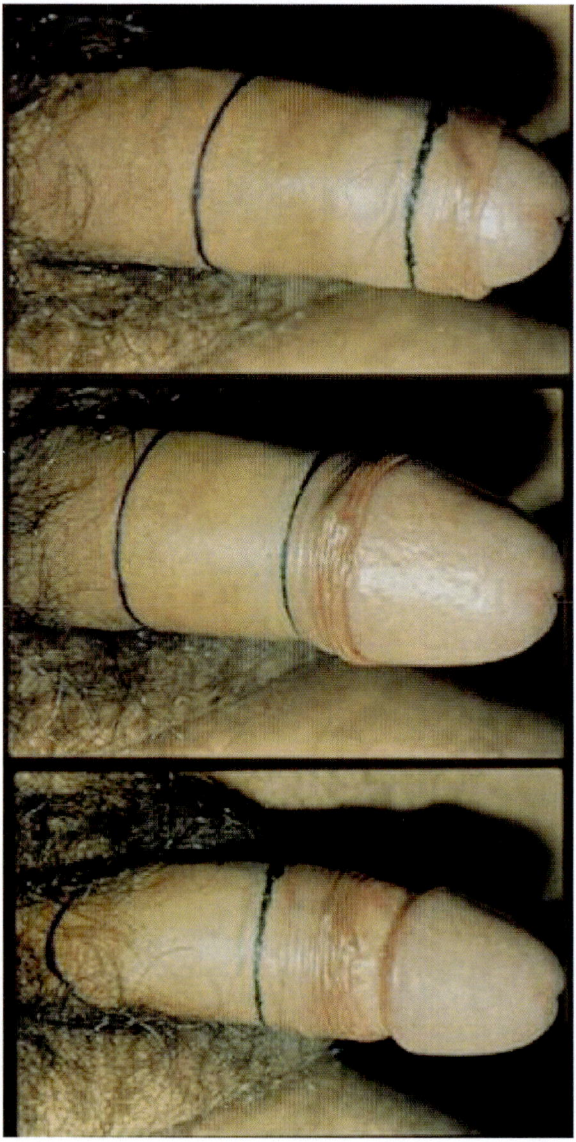

the glans penis to be hidden and protected from injury or unwanted sexual stimulation. The doubled layer of skin including a variable amount of the exposed or outer layer of skin and the inner layer is commonly called the prepuce or foreskin, and this is what is usually removed during a circumcision. Unfortunately, there is no instrument or governing authority which prevents removal of excessive skin or prevents the complications described here.

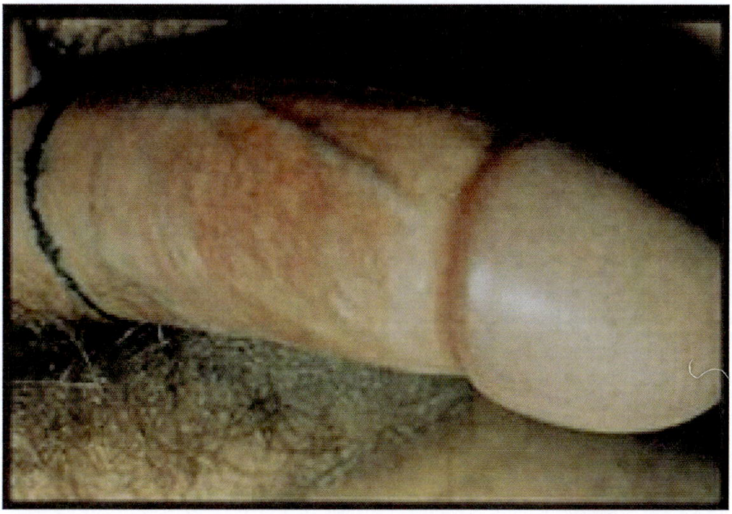

The Supposed Benefits of Circumcision

Since the introduction of circumcision into American society in the nineteenth and early twentieth centuries, a series of reputed benefits has been attributed to the practice by prominent people. Most of these are no longer considered relevant, including supposed prevention or cure of: club foot, insanity (due to masturbation), tuberculosis, blindness, hernia, and others.[2,3] With the passage of time, most of these suppositions have been reduced to absurdities. Within the past generation, the commonest theory to justify male circumcision is a supposition that it might prevent cancer of the female cervix or of the penis. Current research demonstrates that both of these conditions are due to Human Papilloma Virus transmitted by sexual activity. At least in women, this cancer is preventable by the widely recommended vaccine, Gardasil.[4]

Currently, it is proposed that circumcision will prevent urinary tract infections(UTI) and prevent HIV/AIDS.[5,6] The most widely published statistics on UTI indicate that only two percent of male children contract UTI. This is hardly an absolute medical indication for routine circumcision. In any case, a significant number of these infections occur in children who are premature or are in neonatal intensive care units, and are by definition not stable enough for circumcision. A large number of the remainder will be cured by a single course of antibiotics. Children who do not respond to this treatment are often found to have congenital urinary tract anomalies that predispose them to UTI, which cannot be prevented by circumcision.

In the case of HIV/AIDS, this disease was first recognized and became a near epidemic in the United States at a time when infant circumcision of the sexually active US male population approached ninety percent. The world's largest prospective random study on prevention of AIDS was conducted in the United States, and it is conclusively demonstrated that AIDS was/is not prevented by circumcision.[7] AIDS is clearly spread by promiscuous sexual activity with multiple partners, regardless of circumcision status.

Complications of Circumcision

It has been thought that circumcision is a safe procedure with a low incidence of complications, but, in fact, the complications are not documented, and the non-fatal complications show up in the office of pediatricians, family practitioners, and urologists long after the original procedure and seem to have little connection with the original procedure in the popular mind. Complications include skin bridges, urethral fistulas, hemorrhage, infection, gangrene, reduced sexual pleasure, and painful erections.

Aside from the loss of normal sexually related tissue and sexual sensation, it should be emphasized that the pain to which the child is subjected during circumcision is real. The early medical advocates of circumcision in the late nineteenth and early twentieth centuries emphasized that this pain was a part of the intentional aspects of the procedure and was to form a lifelong association between genital function and pain, in order to reduce the pleasure derived from sex, especially during masturbation.

Death of a child during or immediately after a circumcision is rarely reported as a complication of circumcision. Consequently, there are no mortality statistics on circumcision. The death is often stated to be due to hemorrhage, or infection, or some other condition clearly a result of the circumcision as to cause and effect. This lack of awareness results in the incredulous statements of parents and press such as, "What can go wrong with a circumcision?" attributed to the father of a child who died at Rainbow Children's Hospital in Cleveland in October 1998.[8] There are enough sporadic reports of circumcision deaths to make an informal estimate of 200 deaths per year in the United States. Physicians who advocate and perform circumcision have been heard to say that the number of deaths is not significant, even if it approaches figures greater than 200.

This is an illustration of epithelial bridging. This is a not uncommon complication that occurs when the raw surface of the glans of the circumcised penis become an attachment point for the newly formed scar at the circumcision site. It usually is not noticed until years later when it becomes a source of obvious discomfort and disfigurement. It may require anesthesia and surgical separation.

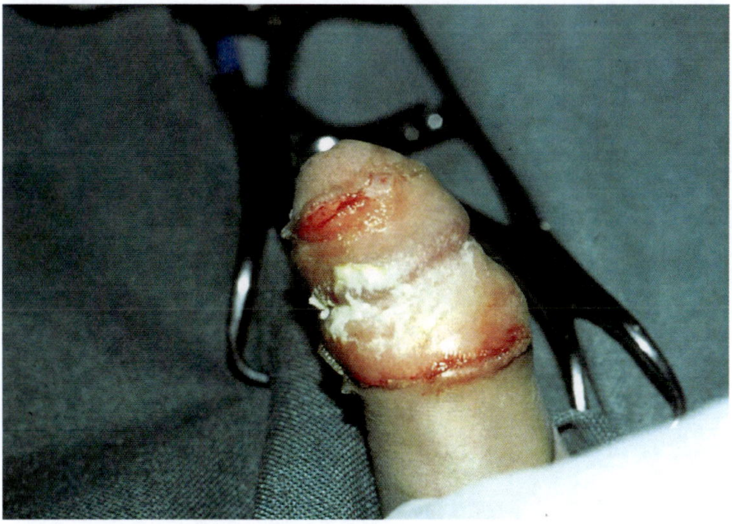

This photo illustrates the immediate post-separation appearance of an epithelial bridge. The accumulated cellular debris after years of entrapment is apparent right after the surgical separation. This material is smegma as found in a circumcised penis.

The following photograph illustrates a urethral fistula and stricture which resulted from a slice across the circumcised infant's glans penis. In addition, it seems that excessive skin was removed from the shaft of the penis. The fistula resulted because the circumcision knife or Gomco clamp included the distal urethra and a portion of the glans penis in the amputated tissue. The natural sequence of healing produced a narrowed exit for the urethra which obstructs the flow of urine and gives this tiny pin-point opening and tiny obstructed flow of urine.

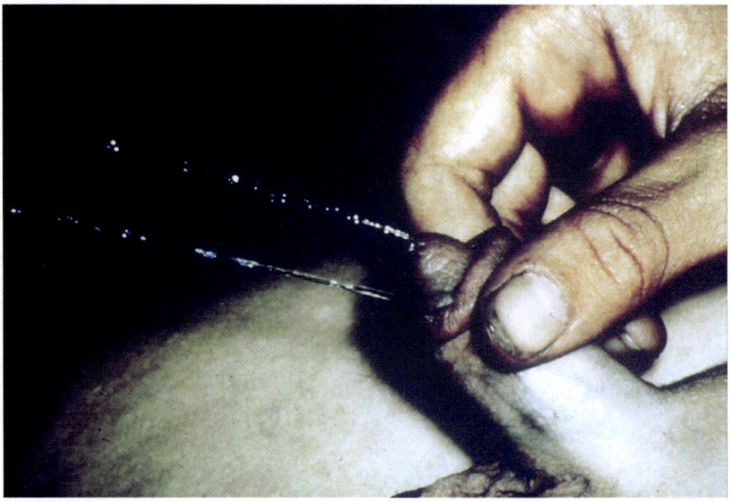

This child is obviously of several years age and has a urethral fistula which probably resulted from a deep cut into the urethra at the time of his infant circumcision. This type of injury requires a skilled urologist expert at urethral repair.

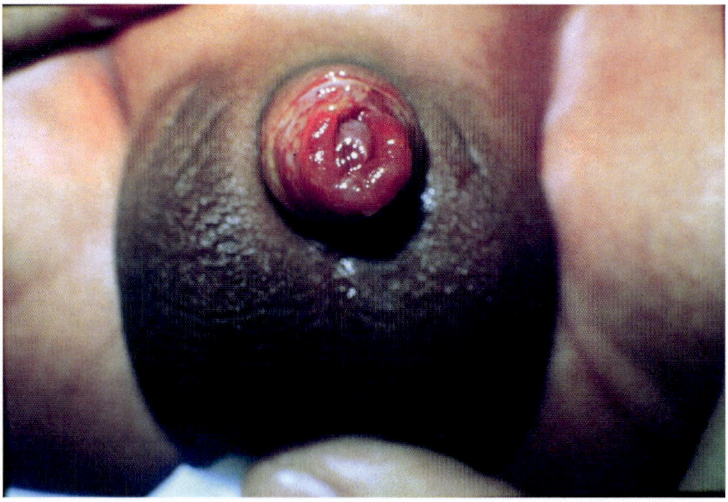

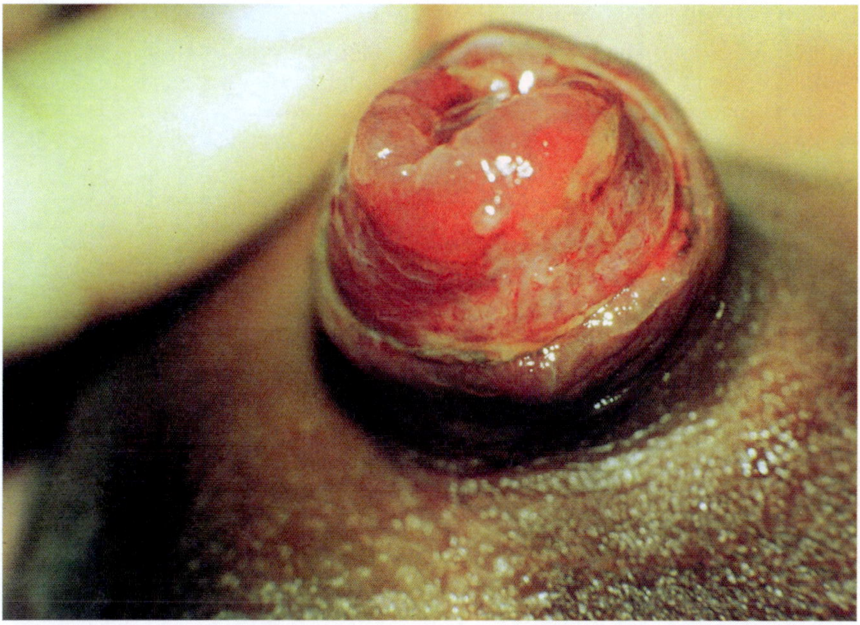

These two pictures above illustrate the separation of a circumcision wound soon after the circumcision was performed. This can occur because most infant circumcisions require no suturing and clots and tissue juices are the adhesive holding the cut edges together. In this child, however, the added problem is that far too much skin of the shaft was removed. Normal erectile activity of the penis caused the shortened skin to be under enough tension to separate the edges. The required treatment here is to suture the edges together. The long-term problem of insufficient penile shaft skin is usually not recognized or managed even though it will be a life-time problem.

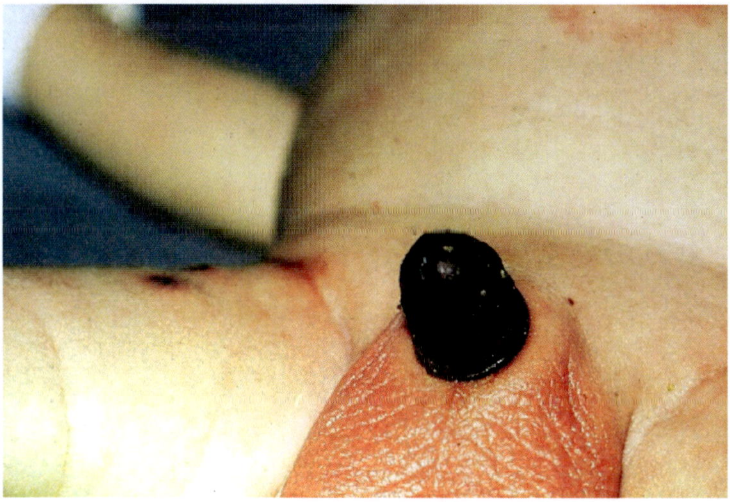

Next to death, this is the most dreaded complication of circumcision. The child depicted above suffered complete necrosis of the penile shaft, almost certainly caused by electrocautery used to control bleeding during the circumcision. This child's normal anatomy can never be replaced by the most skilled surgery. He will be a lifetime genital cripple.

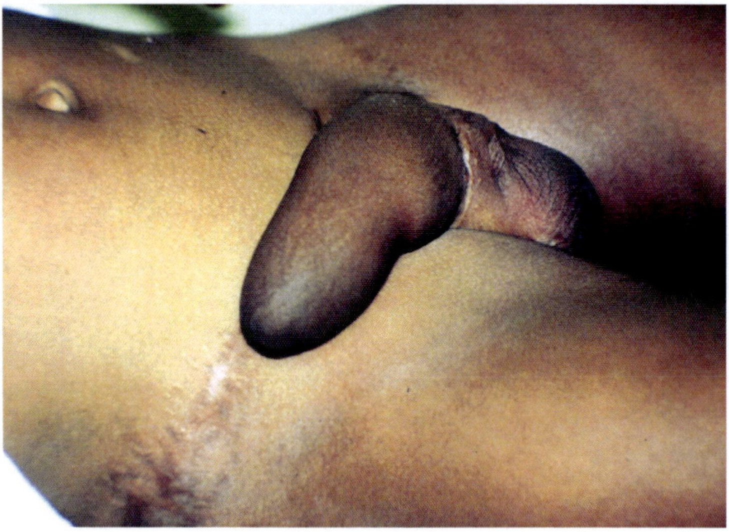

This photo portrays the attempt to reconstruct a penis destroyed by electrocautery during circumcision. Note the scarring on the groin which resulted from taking a full-thickness skin graft in the attempt to recreate the appearance of a normal penis. This pseudo-phallus does not perform the functions of erection, genital sensation, and cannot trigger an orgasm, which are the functions of a normal penis.

Conclusion

In summary, circumcision is a procedure which has repeatedly failed to produce benefits claimed by its advocates. There are no controlled, double blind, prospective studies to prove the claims made for circumcision. The "Possible benefits" claimed for circumcision are illusory. The only certain thing that infant circumcision can prevent is the ability of a pubertal boy to enjoy his normal sexual anatomy and to share it with others when that time comes.

Acknowledgments Photo credits: Due to the difficult nature of this subject material, we have had to rely on multiple sources for the photographic documentation of this article. We are indebted to

Mervyn Landers, a pediatric surgeon and clinical lecturer at the University of Queensland., Boyd Winslow, MD, Steve Scott, John Erickson, James L. Snyder, MD, and others for these materials.

References

1. American Academy of Pediatrics: Task Force on Circumcision. Circumcision policy statement. Pediatrics 1999;103(3):686–693.
2. Darby R. A Surgical Temptation: The Demonization of the Foreskin and the Rise of Circumcision in Britain. Chicago: University Of Chicago Press, 2005.
3. Glick LB. Marked in Your Flesh: Circumcision from Ancient Judea to Modern America. Oxford: Oxford University Press, 2006.
4. Hakim AA, Lin PS, Wilczynski S, Nguyen K, Lynes B, Wakabayashi MT. Indications and efficacy of the human papillomavirus vaccine. Curr Treat Options Oncol 2007 Dec;8(6):393–401.
5. Wiswell TE, Smith FR, Bass JW. Decreased incidence of urinary tract infections in circumcised male infants. Pediatrics 1985;75:901–903.
6. Bailey RC, Plummer FA, Moses S. Male circumcision and HIV prevention: current knowledge and future research directions. Lancet Infect Dis 2001 Nov;1(4):223–231.
7. Millett GA, Ding H, Lauby J, Flores S, Stueve A, Bingham T, Carballo-Dieguez A, Murrill C, Liu KL, Wheeler D, Liau A, Marks G. Circumcision status and HIV infection among Black and Latino men who have sex with men in 3 US cities. J Acquir Immune Defic Syndr 2007 Dec 15;46(5):643–650.
8. http://www.noharmm.org/evansdeath.htm.

Conservative Management of Foreskin Conditions

John Dalton

Abstract Circumcision is seen as a valid, and often necessary, medical treatment even in parts of the world where infant circumcision is rarely performed outside the religious arena. This paper looks at the justifications for circumcision "for medical reasons" in the context of British Medical Association advice that circumcision is unethical and inappropriate where non-invasive treatment is safe and effective. The common clinical indications given for circumcision are reviewed against the evidence for the availability of conservative or non-invasive treatment. This review concludes that the effective treatment preserving the foreskin is available in almost all cases commonly treated by circumcision. The only common condition that may justify circumcision is preputial lichen sclerosus which does not respond to potent topical steroids. New developments suggest that circumcision may perhaps be avoided even in these cases.

Introduction

This paper, from a United Kingdom perspective, looks on the practice of circumcision as a medical treatment and looks critically at the indications given by medical practitioners for performing circumcision as a medical treatment.

Background

The American model of routine infant circumcision is not practiced in the United Kingdom (UK). Parents giving birth to their children in National Health Service (NHS) hospitals are not usually offered a choice of male infant circumcision.

John D Dalton
Researcher and Archiver, NORM-UK, PO Box 71, Stone, Staffordshire, ST15 OSF
www.norm-uk.org, www.genitalautonomy.org

G. Denniston et al. (eds.), *Circumcision and Human Rights*,

Rather the presumption that the normal child has no disease and requires no treatment will usually prevail. Parents of a secular background do not have to make "the circumcision decision" since there is no decision to make.

Circumcision in a medical context, however, is commonly practiced in the UK, since it is regarded as appropriate medical treatment for a plethora of indications, many of which are minor or self-limiting. This accounted for 8,866 circumcisions of boys under the age of 15 in NHS hospitals in England for the statistical year ending in 2003.[1] This implies that 3.1% of English boys will be circumcised for medical reasons by their fifteenth birthday. At older ages the proportion of men circumcised for non-religious reasons will be higher than this, reflecting the fact that any intact man remains at risk for medical circumcision throughout life and the higher incidence of circumcision in earlier years.

It is likely that at least as many circumcisions of non-consenting children take place for religious reasons, although definitive statistics for this are non-existent. Most religiously motivated circumcisions in the UK occur within the Muslim community and Muslim groups have lobbied vigorously for circumcision of children to be provided by the NHS. This has led to the establishment of a number of "circumcision services."[2] While these services are ostensibly to meet religious requirements of the parents, they provide a foot in the door for non-therapeutic circumcision of children on demand. At the high profile circumcision service in Bradford, leaflets soliciting for the service are given to all mothers attending antenatal clinics in the city.[3] Some overall statistics for male circumcision in the UK were provided by the Sexual Attitudes and Lifestyles Survey, published in 1994.[4] This document reported that, overall, 21.8% of all British men are circumcised but that the figure was 12.5% for those aged 16–24 and 32.3% for those aged 45–59. The survey found that white men were the least likely to be circumcised. An update to this survey, published in 2003, reported that 15.8% of British men are circumcised with a peak in age range 40–44. Jews were most likely to be, while Sikhs, Hindus, and Buddhists were least likely.

Several studies conducted in England in recent years have shown that most circumcisions for medical reasons are unnecessary.[1,5–9] This is of concern since, from an ethical standpoint, surgical intervention should be justified by the presence of disease, the unavailability of non-invasive treatment, the effectiveness of the surgical intervention, and the informed consent of the patient.

In the context of male circumcision, the British Medical Association (BMA) has expressed a similar view:

> Unnecessarily invasive procedures should not be used where alternative, less invasive techniques, are equally efficient and available. It is important that doctors keep up to date and ensure that any decisions to undertake an invasive procedure are based on the best available evidence. Therefore, to circumcise for therapeutic reasons where medical research has shown other techniques to be at least as effective and less invasive would be unethical and inappropriate.[10]

That two-thirds of circumcisions of children were unnecessary, since doctors did not understand the natural history of the foreskin, was noted by the House of Commons Health Committee in 1997.[11] HM Government responded, in a Command Paper, that "Surgical interventions should only be performed when clinically necessary, especially in children."[12]

Reasons for Medical Circumcision

The most common indication given for circumcision is phimosis, accounting for 90% of operations.[1,13] Phimosis, however, is an ill-defined term and clear diagnostic criteria are needed if sensible decisions on treatment are to be made.

An as-yet-unpublished document[14] by the British Association of Pediatric Surgeons (BAPS) and British Association of Pediatric Urologists (BAPU) has stated that the indications for include "balanitis xerotica obliterans," recurrent balanoposthitis, and recurrent febrile UTIs where an abnormal urinary tract is present.

I will examine critically the treatment of these conditions, together with other conditions that may be deemed to be indications for circumcision by less erudite practitioners.

Phimosis

The concept of phimosis, from the Greek for muzzling, seems to cover three main concepts:

The non-retractable foreskin of childhood
The foreskin that remains tight or non-retractable into adult life
The foreskin that becomes non-retractable secondary to some other condition such as lichen sclerosus

Most referrals for pediatric circumcision are for nothing more than the normal non-retractable foreskin of childhood.[1, 5–9] Current British teaching on the normal development of the foreskin relies almost solely on Gairdner's paper of 1949.[15] This landmark study, however, is inadequate to inform current practice since it looks only at the development of the foreskin in younger boys and fails to chart the prognosis for school age boys who still have a non-retractable foreskin. The long-term follow-up necessary to chart the development of the foreskin through to adolescent or adult life has been addressed in five more recent studies.[16–20] These studies come from diverse ethnic groups and show a remarkably consistent picture: the normal foreskin may not become retractable until late adolescence. BAPS/BAPU have confirmed that the non-retractable foreskin need not in itself be an indication for treatment below age nineteen years.[14]

As such, it is clear that the histologically normal non-retractable foreskin in a prepubescent child is not a disease and requires no treatment.

When treatment for phimosis is indicated because of the age of the patient, many published studies show that uncomplicated phimosis can be treated by simple stretching techniques.[21,22] An extension of this is to use balloon dilators.[23] The next choice of treatment for phimosis is that of potent topical agents — usually steroids. The evidence on this has been covered by three reviews.[24–26]

Phimosis is also amenable to treatment with plastic correction, conserving the foreskin and removing no tissue.[27–40]

"Ballooning" of the prepuce is not associated with objective measures of urinary obstruction.[41]

Thus, there is overwhelming evidence that circumcision cannot ethically be indicated as a treatment for phimosis where there is no histopathology and the patient is a non-consenting child.

Phimosis due to lichen sclerosus requires more detailed consideration.

Lichen Sclerosus/BXO

So-called balanitis xerotica obliterans (BXO) has been referred to as "true phimosis."[42–45] The medical evidence, however, confirms that BXO is a misnomer for lichen sclerosus.[46–51] It is worth noting that the term balanitis implies inflammation of the glans.

While it is self-evident that circumcision removes lichen sclerosus of the foreskin (Gk: posthe), I am aware of no evidence from randomized controlled trials (RCTs) to show that circumcision is effective for the treatment or prevention of lichen sclerosus of the glans. Several researchers have stated that it is not effective[52–59] or that BXO is associated with the circumcised condition.[60–70]

The standard treatment for lichen sclerosus (LS), in any other context, would be a potent topical steroid.[71] Clinical guidelines for this condition exist, confirming that topical steroids are appropriate treatment.[70]

BAPS states that BXO is an indication for circumcision on the basis that there is no evidence from RCTs to show that topical steroids are effective for treating this condition.[14] However, there is RCT evidence to support the treatment of phimosis associated with lichen sclerosus/BXO. Lindhagen presented a prospective, randomized, double-blind study, although it is unclear from the report whether those who were effectively treated actually had lichen sclerosus.[72] More convincing evidence comes from Kiss and colleagues, who also presented a randomized, placebo controlled double blind study to show the effective treatment of BXO histopathology by mometasone furoate.[73] Jorgensen reported a success rate of 70% for clobetasol dipropionate.[74] While this latter study may be less rigorous than an RCT, there appears to be a relationship between the percentage efficacy and the potency of the steroid. See the table below.

ID	Study	Steroid	% Efficacy
1.	Vincent and MacKinnon (2005)[77]	Hydrocortisone/Triamcynalone	19%
2.	Kiss et al. (2001)[73]	Mometasone furoate	41%
3.	Lindhagen (1996)[72]	Clobetasol propionate	70%
4.	Jorgensen and Svensson (1993)[74]	Clobetasol propionate	70%

Additionally, there is a substantial basis in case reports or uncontrolled trials to support the use of potent topical steroids for penile LS (aka BXO).[50,55,57,61,63,75–81]

Potent topical steroids and circumcision are not the only treatment options for penile lichen sclerosus. Other treatments to be described include: sublesional or intralesional topical steroids,[61–80] carbon-dioxide laser treatment,[49–52,82,83] long-term antibiotic therapy,[84] prepuceplasty with intralesional steroids,[77] and topical tacrolimus.[85,86]

Recurrent Balanitis

I am aware of no reliable evidence that circumcision is effective in reducing recurrences of balanitis.

Recurrent balanitis has been reported to be a form of irritant dermatitis due to excessive hygiene and can be managed by restriction of washing with soap.[87,88] There are existing very detailed clinical guidelines for the management of balanitis.[89] The guidelines do not recommend circumcision for the condition, except in cases of lichen sclerosus where phimosis develops (see above for analysis of this) or the rare condition known as Zoon's balanitis.

While the guidelines for balanitis quote anecdotal evidence that Zoon's balanitis resolves following circumcision, there are reports of the effectiveness of conservative management for this condition either by means of carbon dioxide laser treatment,[90,91] erbium:YAG laser treatment[92] or by topical tacrolimus.[93,94] It, therefore, is not possible to conclude that circumcision can ethically be justified for Zoon's balanitis.

Paraphimosis

BAPS/BAPU do not recommend circumcision for paraphimosis, but many practitioners regard this condition as an indication for "interval" circumcision or even emergency circumcision. There, however, is a plethora of simple conservative treatments for this condition — some of which could be implemented at home by a properly informed patient. A review of this was provided by Little.[95] It is difficult to see how or why circumcision should be considered as a response to paraphimosis.

Prevention of UTIs in Boys with VUR

Perhaps surprisingly, the new BAPS/BAPU document has introduced a new and worrying indication for circumcision: the prevention of UTIs in boys with vesicourethral reflux (VUR). This is of particular concern in the light of a recent RCT of boys having anti-reflux surgery and randomized as to whether they also had a circumcision, which concluded that circumcision was not effective in reducing recurrences of UTI.[96] It would be unfortunate if circumcision were to be accepted as a treatment for this condition after it has already been shown not to work.

Miscellaneous Indications for Circumcision

There is anecdotal evidence about practitioners who will advocate circumcision for a variety of conditions, including lymphedema of the penis or hydrocele. In the absence of any RCT evidence of effectiveness, circumcision for these conditions is highly inappropriate. Minor conditions affecting the foreskin, such as spots, warts, or soreness should be treated as they would be elsewhere on the body.

Discussion

Circumcision "for medical reasons" continues to be a common procedure in the UK, regarded by medical practitioners as a cheap and simple procedure, providing a "quick fix" for a variety of minor problems.

Practitioners performing or recommending circumcision seem oblivious to the potential for harm. Every circumcision removes the mechanical gliding function of the foreskin and also removes the neurological function of the specialized ridged band, which has a possible role in the afferent limb of the ejaculatory reflex.[97] As such, it is perhaps unsurprising that recent studies have reported a reduction of sexual satisfaction following circumcision in 17%,[98] 27%,[99] and 38%[100] of patients.

In addition to the disfigurement and dysfunction that is inherent in the procedure, reported complication rates for circumcision have been as high as 55%.[101] In one NHS series, over 20% of day-case circumcisions required later intervention by the GP for a complication.[102] A realistic rate of significant complications for circumcision lies in the range of 2% to 10%.[103] As such, it is possible that the number of patients experiencing a complication of circumcision equals or exceeds the number of patients with a defensible therapeutic need for the procedure.

As a charity, NORM-UK has received and responded to many letters from men who object to having been circumcised as children and who have suffered decades of anguish over a procedure that was imposed on them without personal consent and, in most cases, without adequate therapeutic need. Children should be given competent treatment, based on what they actually need rather than what someone else wishes.[104]

We also receive many letters from patients or their parents asking for advice on the conservative management of foreskin conditions because they wish to avoid circumcision when their doctors are telling them that there is no alternative. Most of the patients listed for circumcision either have no disease or could be treated non-invasively in less time than they would spend on the waiting list for circumcision. Given the BMA guidance[10] on circumcision, current UK medical practice in this area is unethical.

Overwhelming evidence shows that two-thirds of the circumcisions for phimosis are unnecessary since they are merely treating the normal non-retractable foreskin of childhood. The findings of Lindhagen[72] and Jorgensen[74] suggest that 70% of those cases where lichen sclerosus is present could be treated effectively by topical clobetasol dipropionate. On this basis, a simple but conservative estimate of the fraction of UK medical circumcisions that may be necessary can be made as follows:

$$(100\% - 66\%) \times (100\% - 70\%) = 10\%$$

The recent work of Vincent and MacKinnon reports success in treating preputial lichen sclerosus by means of prepuceplasty combined with intralesional triamcynolone.[77] This implies that even circumcision for preputial lichen sclerosus may be obsolete.

That said, circumcision remains a valid treatment option for a sexually active adult with lichen sclerosus of the foreskin. However, given reports that the condition may resolve spontaneously over a number of years,[105] it would seem inappropriate to consider surgery in childhood while there is a potential for resolution before puberty.

References

1. Cathcart P, Nuttall M, van der Meulen J, Emberton M, Kenny SE. Trends in paediatric circumcision and its complications in England between 1997 and 2003. Br J Surg 2006 July; 93(7):885–90.
2. Pape S. Meeting demand — circumcision. NHS Magazine, Summer 1997.
3. Private communication from David Jackson, Chief Executive of Bradford Hospitals NHS Trust, 1997.
4. Johnson AM, Wadsworth J, Wellings K, Field J, Bradshaw S. Sexual attitudes and lifestyles. Oxford: Blackwell Scientific, 1994 (ISBN: 0-632-03343-6).
5. Griffiths DM, Frank JD. Inappropriate circumcision referrals by GPs. J Royal Soc Med 1992; 85:324–5.
6. Rickwood AMK, Walker J. Is phimosis overdiagnosed and are too many circumcisions performed as a result? Ann Royal Coll Surg Engl 1989; 71:275–7.
7. Williams N, Chell J, Kapila L. Why are children referred for circumcision? (letter). BMJ 1993; 306:28.
8. Farshi Z, Atkinson KR, Squire R. A study of clinical opinion and practice regarding circumcision. Arch Dis Child 2000; 83:393–6.
9. Huntley JS, Bourne MC, Munro FD, Wilson-Storey D. Troubles with the foreskin: one hundred consecutive referrals to paediatric surgeons. J Royal Soc Med 2003; 96(9); 449–51.
10. British Medical Association. The law and ethics of male circumcision: guidance for doctors. London: BMA, 2006.
11. Hospital services for children and young people, House of Commons Health Committee, 16 March 1997.
12. Cm 3793. Government response to the reports of the health committee on health services for children and young people, Session 1996–1997.
13. NORM-UK Survey of Health Authorities (unpublished), NORM-UK, 1999.
14. Management of Foreskin Conditions (unpublished), British Association of Paediatric Surgeons, June 2006.
15. Gairdner D. The Fate of the foreskin. Br Med J 1949; 2:1433–7.

16. Øster J. Further fate of the foreskin. Arch Dis Child 1968; 43:200–3.
17. Morales Concepcion JC, Cordies Jackson E, Guerra Rodriguez M, Mora Casaco B, Morales Aranegui A, Gonzalez Fernandez P. [Should circumcision be performed in childhood?]. Arch Esp Urol 2002; 55(7):807–11.
18. Kayaba H, Tamura H, Kitajima S, Fujiwara Y, Kato T, Kato T. Analysis of shape and retractability of the prepuce in 603 Japanese boys. J Urol 1996; 156:1813–5.
19. Agarwal A, Mohta A, Anand RK. Preputial retraction in children. J Indian Assoc Pediatr Surg 2005; 10(2):89–91.
20. Thorvaldsen MA, Meyhoff HH. Patologisk eller fysiologisk fimose? [Pathological or physiological phimosis?]. Ugeskr Laeger 2005; 167(17):1858–62.
21. Beaugé MD. The causes of adolescent phimosis. Br J Sex Med 1997; 24(5):26.
22. Zampieri N, Corroppolo M, Camoglio FS, Giacomello L, Ottolenghi A. Phimosis: stretching methods with or without application of topical steroids? J Pediatr 2005 November; 147(5):705–6.
23. Ying H, Xiu-hua Z. Balloon dilation treatment for phimosis in boys. Chinese Med J 1991; 104:491–3.
24. Dewan PA, Tieu HC, Chieng BS. Phimosis: is circumcision necessary? J Paediatr Child Health 1996; 32(4):285–9.
25. Berdeu D, Sauze L, Ha-Vinh P, Blum-Boisgard C. Cost-effectiveness of treatments for phimosis: a comparison of surgical and medicinal approaches and their economic effect. BJU Int 2001; 87:239–44.
26. Van Howe RS. Cost-effective treatment of phimosis. Pediatrics 1998; 102:E43.
27. Hoffman S, Metz P, Ebbehoj J. A new operation for phimosis: prepuce saving technique with multiple Y-V plasties. Br J Urol 1984; 56:319–21.
28. de Castella H. Prepuceplasty: an alternative to circumcision. Ann Royal Coll Surg Engl 1994; 76(4):257–8.
29. Cuckow PM, Rix G, Mouriquand PDE. Preputial plasty: a good alternative to circumcision. J Pediatr Surg 1994; 29(4):561–3.
30. Cuckow P, Mouriquand P. Saving The Normal Foreskin (Letter). Br Med J. 1993; 306:459-60.
31. Emmett AJJ. Z Plasty reconstruction for preputial stenosis: a surgical alternative to circumcision. Aust Paediatr J 1992; 18:219–20.
32. Fleet MS, Venyo AKG, Rangecroft L. Dorsal relieving incision for the non-retractile foreskin. J Royal Coll Surg Edinb 1995; 40:243–5.
33. Parkash S. Phimosis and its plastic correction. J Ind Med Assoc 1972; 58:389.
34. Cicia S, Florio G. Postectomy for phimosis: 5 year experience. Chir Ital 2000; 52(6):733–6.
35. W halin N. "Triple incision plasty" a convenient procedure for preputial relief. Scand J Urol Nephrol 1992; 26:107–10.
36. Holmlund D. Dorsal incision of the prepuce and closure with dexon in patients with phimosis. Scand J Urol Nephrol 1973; 7:97–89.
37. Gaetini AM. [Preputial plastic surgery creating a two-way sliding rim in the treatment of phimosis]. Minerva Pediatr 1984; 36(18):905–7.
38. Lane TM, South LM. Lateral prepuceplasty for phimosis. J Royal Coll Surg Edinb 1999; 44:310–2.
39. Codega G, Guizzardi D, Di Giuseppe P, Fassi P. [Helicoid plasty in the treatment of phimosis]. Minerva Chir 1983; 38(22):1903–7.
40. Saxena AK, Schaarschmidt K, Reich A,Willital GH. Non-retractile foreskin: a single center 13-year experience. Int Surg 2000; 85(2):180–3.
41. Babu R, Harrison SK, Hutton KA. Ballooning of the foreskin and physiological phimosis: is there any objective evidence of obstructed voiding? BJU Int 2004; 94(3):384–7.
42. Rickwood AMK. Medical indications for circumcision. BJU Int 1999; 83(Suppl 1):45–51.
43. Rickwood AMK, Kenny E, Donnell C. Towards evidence based circumcision of English boys: survey of trends in practice. Br Med J 2000; 321:792–3.

44. Rickwood AMK, Hemalatha G, Batcup G, Spitz L. Phimosis in boys. Br J Urol 1980; 52:147–50.
45. Shankar KR, Rickwood AMK. The Incidence of phimosis in boys. BJU Int 1999; 84:101–2.
46. Hinchliffe SA, Ciftci AO, Khine MM, Rickwood AMK, Ashwood J, McGill F, Clapham EM, van Velzen D. Composition of the inflammatory infiltrate in pediatric penile lichen sclerosus et atrophicus (balanitis xerotica obliterans): a prospective, comparative immunophenotyping study. Pediatr Pathol 1994; 14:223–33.
47. Laymon CW, Freeman C. Relationship of balanitis xerotica obliterans to lichen sclerosus et atrophicus. Arch Derm Syph 1994; 49:57–9.
48. Potter B. Balanitis xerotica obliterans manifesting on the stump of amputated penis. Arch Dermatol 1959; 79:473.
49. Ratz JL. Carbon dioxide laser treatment of balanitis xerotica obliterans. J Am Acad Dermatol 1984; 10:925–8.
50. Pasieczny TAH. The treatment of balanitis xerotica obliterans with testosterone propionate ointment. Acta Derm Venerol (stockh) 1977; 57:275–7.
51. Clark C, Huntley JS, Munro FD, Wilson-Storey D. Managing the paediatric foreskin. Practitioner 2004; 248(1665):888–94.
52. Rosemberg SK. Carbon dioxide laser treatment of external genital lesion. Urology 1985; 26:555–8.
53. Datta C, Dutta SR, Chaudhuri A. Histopathological and immunological studies in a cohort of balanitis xerotica obliterans. J Ind Med Assoc 1993; 91:146–8.
54. Pride HB, Miller F, Tyler W A. Penile squamous cell carcinoma arising from balanitis obliterans. J Am Acad Dermatol 1993; 29:469–73.
55. Herschorn S, Colapinto V. Balanitis xerotica obliterans involving anterior urethra. Urology 1979; 14:592–6.
56. Garat JM, Checile G, Algaba F, Santaularia JM. Balanitis xerotica obliterans in children. J Urol 1988; 136:136–7.
57. Hrebinko RL. Circumferential laser vaporization for severe meatal stenosis secondary to balanitis xerotica obliterans. J Urol 1996; 156:1735–6.
58. Jorgensen ET, Svensson A. Problems with the penis and prepuce in children. Lichen sclerosus should be treated with corticosteroids to reduce need for surgery (letter). Br Med J 1996; 313(7058):692.
59. Das S, Tunuguntla HS. Balanitis xerotica obliterans — a review. World J Urol 2000; 18(6):382–7.
60. Stuehmer A. Balanitis xerotica obliterans (post operationem) und ihre beziehungen zur "kraurosis glandis et praeputii penis." Arch Derm Syph 1928; 156:613–23.
61. Poynter JH, Levy J. Balanitis xerotica obliterans: effective treatment with topical and sublesional corticosteroids. Br J Urol 1967; 39:420–5.
62. Zungri E, Chechile G, Algaba F, Mallo N. Balanitis xerotica obliterans: surgical treatment. Eur Urol 1988; 14:160–2.
63. Bale PM, Lochhead A, Martin HC, Gollow I. Balanitis xerotica obliterans in children. Pediatr Pathol 1987; 7:617–27.
64. Weigand DA. Lichen sclerosus et atrophicus, multiple displastic keratosis and squamous cell carcinoma of the glans penis. J Dermatol Surg Oncol 1980; 6:45–50.
65. Freeman C, Laymon CW. Balanitis xerotica obliterans. Arch Derm Syph 1941; 44:547–61.
66. Franks AG. Balanitis xerotica obliterans. J Urol 1946; 56:243.
67. Potter B. Balanitis xerotica obliterans manifesting on the stump of amputated penis. Arch Dermatol 1959; 79:473.
68. Harova G, Özkara SK, Gürbüz Y, Çulha M, Yumbul Z. A case of balanitis xerotica obliterans with koilocytosis. Turk J Dermatopathol 1998; 7(3–4).
69. Ledwig PA, Weigand DA. Late circumcision and lichen sclerosus et atrophicus of the penis. J Am Acad Dermatol 1989; 20(2 Pt1):211–4.

70. Neill SM, Tatnall FM, Cox NH. Guidelines for the management of lichen sclerosus. Br J Dermatol 2002; 147:640–9.
71. Ledingham JGG, Warrell DA (Eds). Concise Oxford textbook of medicine. OUP, 2000 (ISBN: 0–19–262870–4). Oxford University Press is located in Oxford, UK.
72. Lindhagen T. Topical clobetasol propionate compared with placebo in the treatment of the unretractable foreskin. Eur J Surg 1996; 162:969.
73. Kiss A, Csontai A, Pirot L, Nyirady P, Merksz M, Kiraly L. The response of balanitis xerotica obliterans to local steroid application compared with placebo in children. J Urol 2001; 165(1):219–20.
74. Jorgensen ET, Svensson A. The Treatment of phimosis in boys, with a potent topical steroid (clobetasol propionate 0.05%) Cream. Acta Derm Venerol (stockh) 1993; 73(1):55–6.
75. Dahlman-Ghozlan K, Hedbald MA, Von Krogh G. Penile lichen sclerosus et atrophicus treated with clobetasol diproprionate 0:05% cream: a retrospective clinical and histopathalogical study. J Am Acad Dermatol 1999; 40(3):451–7.
76. Neuhaus IM, Skidmore RA. Balanitis xerotica obliterans and its differential diagnosis. J Am Board Fam Pract 1999; 12(6):473–6.
77. Vincent MV, MacKinnon E. The response of clinical balanitis xerotica obliterans to the application of topical steroid-based creams. J Pediatr Surg 2005; 40(4):709–12.
78. Catterall RD, Oats JK. Treatment of balanitis xerotica obliterans with hydrocortisone injections. Br J Vener Dis 1962; 38:75–8.
79. Campus GV, Ena P, Scuderi N. Surgical treatment of balanitis xerotica obliterans. Plast Reconstr Surg 1984; 73(4):652–7.
80. Uemura S, Hutson JM, Woodward AA, Kelly JH, Chow CW. Balanitis xerotica obliterans with urethral stricture after hypospadias repair. Pediatr Surg Int 2000; 16(1–2):144–5.
81. Riddell L, Edwards A, Sherrard J. Clinical features of lichen sclerosus in men attending a department of genitourinary medicine. Sex Transm Infect 2000; 76:311–3.
82. Kartamaa M, Reitamo S. Treatment of lichen sclerosus with carbon dioxide laser vaporization. Br J Dermatol 1997; 136:365–9.
83. Windahl T, Hellsten S. Carbon dioxide laser treatment of lichen sclerosus et atrophicus. J Urol 1993; 150:868–70.
84. Shelley W B, Shelley ED, Grunenwald MA, Anders TJ, Ramnath A. Long-term antibiotic therapy for balanitis xerotica obliterans. J Am Acad Dermatol 1999; 40:69–72.
85. Assmann T, Becker-Wegerich P, Grewe M, Megahed M, Ruzicka T. Tacrolimus ointment for the treatment of vulvar lichen sclerosus. J Am Acad Dermatol 2003; 48(6):935–7.
86. Meykadeh N, Hengge UR. Topische Immunmodulation in der Dermatologie [Topical immunomodulators in dermatology]. Hautarzt 2003; 54(7):641–61.
87. Birley HDL, Walker MM, Luzzi GA, Bell R, Taylor-Robinson D, Byrne M, Renton AM. Clinical features and management of recurrent balanitis: association with atopy and genital washing. Genitourin Med 1993; 69:400–3.
88. Kohn FM. [Article in German]. MMW Fortschr Med 2002; 144(12):30–5.
89. No author identified. National guideline for the management of balanitis. Sex Transm Infect 1999; 75(Suppl 1):S85–8.
90. Retamar RA, Kien MC, Chouela EN. Zoon's balanitis: presentation of 15 patients, five treated with a carbon dioxide laser. Int J Dermatol 2003; 42(4):305–7.
91. Baldwin HE, Geronemus RG. The treatment of Zoon's balanitis with the carbon dioxide laser. J Dermatol Surg Oncol 1989; 15:491–4. Not held. Shows CO2 laser ablation to be an effective alternative to circumcision for treatment of ZB.
92. Albertini JG, Holck DE, Farley MF. Zoon's balanitis treated with Erbium: YAG laser ablation. Lasers Surg Med 2002; 30(2):123–6.
93. Hernandez-Machin B, Hernando LB, Marrero OB, Hernandez B. Plasma cell balanitis of Zoon treated successfully with topical tacrolimus. Clin Exp Dermatol 2005; 30(5):588–9.
94. Santos-Juanes J, Sanchez del Rio J, Galache C, Soto J. Topical tacrolimus: an effective therapy for Zoon balanitis (letter). Arch Dermatol 2004; 140(12):1538–9.

95. Little B, White M. Treatment options for paraphimosis. Int J Clin Pract 2005; 55(9):591–3.
96. Kwak C, Oh SJ, Lee A, Choi H. Effect of circumcision on urinary tract infection after successful antireflux surgery. BJU Int 2004; 94(4):627–9.
97. Taylor JR, Lockwood AP, Taylor AJ. The prepuce: specialized mucosa of the penis and its loss to circumcision. BJU Int 1996; 77:291–5.
98. Masood S, Patel HR, Himpson RC, Palmer JH, Mufti GR, Sheriff MK. Penile sensitivity and sexual satisfaction after circumcision: are we informing men correctly? Urol Int 2005; 75(1):62–6.
99. Coursey JW, Morey AF, McAninch JW, Summerton DJ, Secrest C, White P, Miller K, Pieczonka C, Hochberg D, Armenakas N. Erectile function after anterior urethroplasty. J Urol 2001; 166(6):2273–6.
100. Fink KS, Carson CC, DeVellis RF. Adult circumcision outcomes study: effect on erectile function, penile sensitivity, sexual activity and satisfaction. J Urol 2002; 167(5):2113–6.
101. Patel H. The problem of routine circumcision. Can Med Assoc J 1966; 95:576–581.
102. de la Hunt MN. Paediatric day care surgery: a hidden burden for primary care? Ann R Coll Surg Engl 1999; 81:179–82.
103. Williams N, Kapila L. Complications of circumcision. Br J Surg 1993; 80:1231–6.
104. American Academy of Pediatrics Committee on Bioethics. Informed consent, parental permission, and assent in pediatric practice. Pediatrics 1995; 95(2):314–7.
105. Helm KF, Gibson LE, Muller SA. Lichen sclerosus et atrophicus in children and young adults. Pediatr Dermatol 1991; 8:97.

Methicillin-Resistant *Staphylococcus aureus*: An Emerging Risk for Circumcised Boys

George Hill

Abstract *Staphylococcus aureus* has been treated with antibiotic regimens for more than six decades. The organism has shown a remarkable ability to evolve resistance to commonly used antibiotics. The resistant strains commonly are called methicillin-resistant *Staphylococcus aureus* (MRSA). The antibiotic-resistant varieties, which commonly had been found in hospitals, recently have entered the community, where their prevalence, in many areas, are reaching epidemic proportions. MRSA is carried on the skin and in the nares of healthy people. Healthcare workers and parents may colonize newborn infants with MRSA. Any open wound, including a circumcision wound, increases the risk of infection. *Staphylococcus aureus* commonly causes skin infections, but it may also cause fulminating necrolytic pneumonia, meningitis, necrotizing fasciitis, and other life-threatening systemic infections. The community-associated strains (CA-MRSA) have developed new virulence factors not previously seen in hospital-associated strains (HA-MRSA). Treatment should be aggressive and immediate, but still the death rate is high. Avoidance of non-therapeutic circumcision is indicated.

Introduction

The report by the Centers for Disease Control (CDC) of the deaths of four children in Minnesota and North Dakota from fulminant CA-MRSA in the late 1990s served to wake up the medical community to the dangers of this emerging pathogen.[1] Since that time there has been extensive investigation of this virulent new pathogen. This review examines the risk that the presence of the MRSA pathogen in the community poses to newly circumcised boys.

Staphylococcus aureus

Staphylococcus aureus is a Gram-positive spherical bacteria that occurs in microscopic clusters resembling grapes.[2] *S. aureus* displays a golden color under the microscope, from which it derives its name.[2]

Vice-President for Bioethics and Medical Science, Doctors Opposing Circumcision, Seattle, Washington 98107–4137, USA

S. aureus commonly causes skin infections such as bullous impetigo, furuncles, staphylococcal scalded skin syndrome, and pyoderma.[2,3] *S. aureus* also may cause life-threatening systemic infections such as necrotizing pneumonia, necrotizing fasciitis, and meningitis.[2,3] In addition, *S. aureus* may cause deep-seated infections such as osteomylitis and endocarditis.[2,3]

Most strains of *S. aureus* produce the enzyme coagulase that causes clotting of the blood.[2] Most strains of *S. aureus* also produce tissue-destroying exotoxins, which may cause skin exfoliation, emesis, and toxic shock.[2,3]

S. aureus traditionally has inhabited hospitals and long has been a major cause of nosocomial infection of surgical wounds and indwelling medical devices.[2] *S. aureus* has been reported in hospital nurseries since 1889.[4]

Types of *Staphylococcus aureus*

S. aureus has a remarkable ability to develop resistance to antibiotics.[3,5] Penicillin was introduced in 1943. Pryles reported in 1958 that penicillin had lost much of its effectiveness against *S. aureus*.[5] *S. aureus* has continued to develop resistance to later antibiotics and, today, strains resistant to methicillin, erythromycin, and vancomycin exist,[3,6] all of which are usually grouped under the name, methicillin-resistant *Staphylococcus aureus* (MRSA).

MRSA

MRSA has been associated with hospitals (HA-MRSA) for decades[6] and has traditionally been viewed as a nosocomial infection, which posed an additional risk for patients, especially surgical patients, who were likely to be infected.

MRSA now has escaped from hospitals into the community and is a worldwide problem.[4] Community-associated MRSA (CA-MRSA) is now found in many areas of the world,[6,7] including Canada,[8,9] Australia,[10] New Zealand,[11] the United Kingdom,[12] and the United States.[13]

CA-MRSA

CA-MRSA has acquired new genetic material that differs from that of HA-MRSA.[3,7] Nineteen new genes have been reported.[14] This genetic material renders CA-MRSA more resistant to antibiotics, and more virulent. The most important difference probably is the Panton-Valentine leukocidin-producing gene (PVL) producing gene (PVL).[3,6–8] The PVL gene induces tissue necrosis and leukocyte destruction.[3,7] Abscesses are a striking feature of infection with MRSA containing the PVL gene.[3,4]

CA-MRSA provides a unique combination of increased infectivity, resistance, and virulence that poses a real challenge to healthcare providers.[6,7]

Mortality

There is a substantially increased mortality in patients infected with MRSA. Cosgrove et al. compared deaths from methicillin-sensitive *Staphylococcus Aureus* (MSSA) and MRSA and concluded "that bacteremia due to methicillin-resistant *S. aureus* is associated with increased mortality compared with MSA bacteremia."[15] Healy et al. studied mortality due to CA-MRSA in a neonatal intensive care unit and reported a death rate of 38%.[16] Noskin et al. studied data from the National Inpatient Sample Database and concluded that patients with Staphylococcus infections had

> 3 times the length of hospital stay (14.3 vs 4.5 days; *P* <.001), 3 times the total charges ($48 824 vs $14 141; *P* < .001), and 5 times the risk of in-hospital death (11.2% vs 2.3%; *P* < .001) than inpatients without this infection.[17]

Wyllie et al. studied death rates among patients with *Staphylococcus aureus* infection in Oxfordshire. They reported a death rate of 29% among such patients. MRSA contributed to an increase in the number of patients with Staphylococcal infection.[18] Melzer et al. studied death rates among British patients. They reported:

> The proportion of patients whose death was attributable to methicillin-resistant *S. aureus* (MRSA) was significantly higher than that for methicillin-susceptible *S. aureus* (MSSA) (11.8% vs. 5.1%; ...[19]

Isaacs et al. compared death rates of newborn babies infected with MRSA to the death rate of newborn babies infected with MSSA. They reported:

> The mortality of MRSA sepsis was 24.6% compared with 9.9% for MSSA infections. The mortality of early onset MSSA sepsis, however, was 39% (seven of 18) compared with 7.3% of late onset MSSA infection presenting more than two days after birth.[4]

Risk Factors

CA-MRSA increasingly is displacing other varieties of *Staphylococcus aureus* in the community.[7] Healthy persons may be carriers of CA-MRSA on their skin or in their noses. Patients, healthcare workers, and parents who are carriers may introduce CA-MRSA into the hospital setting. Risk factors for infection include crowded conditions and skin-to-skin contact,[4] which are found in newborn nurseries.

Circumcision of the Newborn as a Risk Factor

Neonatal immune systems are less well developed and function more poorly than in other population groups.[20,21] Males are at greater risk of staphylococcal infection than females.[24–26] Newborns, therefore, carefully should be protected from infection.

There is no doubt that male neonatal circumcision is a risk factor for contraction of *S. aureus*, including MRSA, infection. Wiswell et al. reported that 11% of boys are colonized with *Staphylococcus aureus* within 24 hours after circumcision.[22] Wiswell et al. report *Staphylococcus aureus* is found more frequently in circumcised boys at two weeks of age.[23] Neonatal circumcision creates an open wound on the penis. The infection rate is not dependent upon the type of circumcision device used.[22] Invasive or surgical procedures increase risk of MRSA infection.[3,22,26] According to Bratu et al.

> In the pediatric population, risk factors associated with MRSA infections include premature birth or low birth weight, chronic underlying diseases, prolonged hospitalization, invasive or surgical procedures, indwelling catheters, and prolonged use of antimicrobial agents.[26]

Staphylococcus aureus has caused post-circumcision necrotizing pneumonia,[24,26,27] neonatal septicemia,[28] staphylococcal scalded skin syndrome,[29] and, in combination with other pathogens, necrotizing fasciitis,[30,31] and staphylococcal pyoderma.[32]

Although any infant may be colonized and infected with *Staphylococcus aureus*, studies show that circumcised boys contract infection at a much greater rate. Thomson et al. reported that the infection rate among circumcised boys was twice the rate of infection among non-circumcised boys.[25] Curran & Al-Salihi reported that, in one hospital in New Jersey, boys had 5.5 times more Staphylococcal Scalded Skin Syndrome (SSSS) general exfoliative disease than girls.[33] (In New Jersey in 1980, nearly 100% of the male infants would have been circumcised, since neonatal circumcision then was considered a "routine" procedure.) Stranko et al. reported that staphylococcal impetigo occurred only in circumcised boys at the Geisinger Medical Center, with no cases reported in girls.[34] Enzenauer et al. reported that newborn circumcised males in the newborn nursery had twice the incidence of staphylococcal colonization and pyoderma as non-circumcised males, and commented:

Circumcision, by its very nature, requires more staff-patient "hands-on" contact. The infants are all lined up and their stomachs lavaged clear in preparation for the procedure. The circumcisions are done daily, as a group, in a small area, using reusable circumcision restraints.

Postoperatively, there is also more handling of the diaper area in caring for the fresh, hemorrhagic wound.[35]

This is equally applicable to the CA-MRSA strains of *Staphylococcus aureus*.

Several outbreaks of CA-MRSA in hospital nurseries have already been reported. Zafar et al. reported an outbreak in Virginia and cited evidence of 25 outbreaks of *Staphylococcus aureus* in newborn nurseries, of which three were MRSA.[36] Saiman et al. reported an outbreak of CA-MRSA in a New York hospital nursery.[37] Nambiar et al. reported an outbreak of MRSA in the Children's National Medical Center.[38] Davies et al. reported an outbreak in a special care baby unit.[39] Bratu et al. reported an outbreak of CA-MRSA at an unnamed hospital.[26] Reboli et al. reported an outbreak of MRSA in a neonatal intensive care unit.[40]

Newborn boys typically are discharged from hospital soon after their circumcision and while the circumcision wound is still open. They are vulnerable to CA-MRSA infection at home.[24]

Even after the wound is closed, circumcised boys carry more *Staphylococcus aureus* in their urethras than do intact boys.[22,41] Moreover, (as noted above) carriage of the organism is found more frequently among circumcised than among non-circumcised boys at two weeks of age,[23] so they may still be more vulnerable to infection.

Treatment

Although the CA-MRSA epidemic has been developing for more than 8 years,[1] the American Academy of Pediatrics has no policy as of August 2006 with regard to prevention and treatment of CA-MRSA infection in infants and children. The only policy statement, as of August 2006, concerned infection control in pediatrician's offices and the avoidance of lawsuits,[42] which is neither relevant nor helpful.

Treatment modalities for this emerging pathogen are undergoing constant reappraisal so this paper can only discuss treatment in general terms. Grayson describes a "treatment triangle": wound culture, antibiotic therapy, and surgical incision and drainage of abcesses.[43]

CA-MRSA is a potentially life-threatening fulminating infection,[1] so treatment of suspected CA-MRSA infection should start immediately and be carried out aggressively.

Suspected CA-MRSA should be cultured.[43,44] Patients with severe or systemic infections should be hospitalized and managed with parenteral anti-microbial therapy.[43–45] Antibiotics effective against most CA-MRSA include vancomycin, clindamycin, linezolid, and minocycline.[3,43–45]

Abscesses should be drained.[43–45] Staphylococcal necrotizing fasciitis requires aggressive surgical debridement of infected tissue if the patient is to survive.[30,31]

Discussion

This paper shows that

- HA-MRSA has escaped from hospitals and has entered the community.
- CA-MRSA is an emerging pathogen that now is present in epidemic proportions in the community.
- CA-MRSA has acquired virulent new characteristics.
- The mortality rate is high.
- Children die from fulminating CA-MRSA.
- CA-MRSA has reentered hospitals and frequently is found in newborn nurseries.
- Boys are at greater risk than girls for infection with CA-MRSA.
- Newly circumcised boys are at substantially greater risk of infection with CA-MRSA than intact boys.
- Circumcised boys may be at greater risk even after the wound has healed.

The *Circumcision Policy Statement* provided by the American Academy of Pediatrics[46] was drafted in July 1998, approved in November 1998, and published in March 1999. It predates the general recognition of the danger of CA-MRSA, which did not occur until the Centers for Disease Control and Prevention published a warning on September 22, 1999.[1]

That *Circumcision Policy Statement* found little if any prophylactic value to non-therapeutic neonatal circumcision, but it did not find sufficient adverse effects to prohibit its performance.[46] The guidance offered by that document is not relevant to today's situation.

Circumcision on infection rates. Isaacs et al.[4] failed to record the circumcision status of the males in their study.[47,48] Fortunov et al.[24] failed to report the circumcision rates of those male infants infected and not infected. One supposes that cultural blindness may have prevented these researchers from being aware of the obvious. In spite of these regrettable lapses, there is abundant evidence of the risks posed by male non-therapeutic circumcision.

Conclusion

Male non-therapeutic circumcision sharply increases the risk of infant boys being infected with life-threatening CA-MRSA both in hospital and after leaving the hospital and entering into the community. The American Academy of Pediatrics has not adequately addressed this issue. This shifts the balance between reward and risk sharply toward risk and provides a compelling contraindication to medically-unnecessary non-therapeutic child circumcision. It would be appropriate for hospitals and individual doctors to decline to perform this outmoded operation, even at parental request. Parents should be warned of the risk of CA-MRSA infection posed by any invasive operation, including neonatal and child circumcision.

References

1. Centers for Disease Control and Prevention. Four pediatric deaths from community-acquired methicillin-resistant Staphylococcus aureus — Minnesota and North Dakota, 1997–1999. JAMA 1999;282:1123–5.
2. Todar K. Todar's Online Textbook of Bacteriology. Madison: Department of Bacteriology, University of Wisconsin, 2005. Available at: http://textbookofbacteriology.net/staph.html.
3. Zetola N, Francis JS, Nuermberger EL, Bishai WR. Community-acquired methicillin-resistant Staphylococcus aureus: an emerging threat. Lancet Infect Dis 2005;5(5):275–86.
4. Isaacs D, Fraser S, Hogg G, Li HY. Staphylococcus aureus infections in Australasian neonatal nurseries. Arch Dis Child Fetal Neonatal Ed 2004;89:F331–5.
5. Pryles CV. Staphylococcal pneumonia in infancy and childhood. Pediatrics 1958;21(4):609–23.

---

*Postscript. Two significant articles that fully confirm the findings reported in this paper have been published after this paper was presented to the Ninth International Symposium on Circumcision, Genital Integrity, and Human Rights that met at the University of Washington, Seattle, 24–26 August 2006 on 25 August 2006.[48, 49]

6. Foster T (1996). Staphylococcus. In: Barron S, et al., eds. Barron's Medical Microbiology, 4th edition, University of Texas Medical Branch.
7. Chambers HF. Community-associated MRSA — resistance and virulence converge. New Engl J Med 2005;352(14):1485–7.
8. Allen UD. Public health implications of MRSA in Canada. Can Med Assoc J 2006;175(2):145–6.
9. Gilbert M, MacDonald J, Gregson D, et al. Outbreak in Alberta of community-acquired (USA300) methicillin-resistant Staphylococcus aureus in people with a history of drug use, homelessness or incarceration. Can Med Assoc J 2006;175(2):149–54.
10. Collignon P, Gosbell I, Vickery A, et al. Community-acquired methicillin-resistant Staphylococcus aureus in Australia. Australian Group on Antimicrobial Resistance [letter]. Lancet 1998;352:145–6.
11. Smith JM, Cook GM. A decade of community MRSA in New Zealand. Epidemiol Infect 2005;133(5):899–904.
12. Jeyaratnam D, Reid C, Kearns A, Klein J. Community associated MRSA: an alert to paediatricians. Arch Dis Child 2006;91(6):511–2.
13. Dietrich DW, Auld DB, Mermel LA. Community-acquired methicillin-resistant Staphylococcus aureus in Southern New England children. Pediatrics 2004;113:e347–52.
14. Baba T, Takechi F, Kuroda M, et al. Genome and virulence determinants of high-virulence community acquired MRSA. Lancet 2002;359:1919–27.
15. Cosgrove SE, Sakoulas G, Perencevich EN, et al. Comparison of mortality associated with methicillin-resistant and methicillin-susceptible Staphylococcus aureus bacteremia: a meta-analysis. Clin Infect Dis 2003;36:53–9.
16. Healy CM, Hulten KG, Palazzi DL, et al. Emergence of new strains of methicillin-resistant Staphylococcus aureus in a neonatal intensive care unit. Clin Infect Dis 2004;39:1460–6.
17. Noskin GA, Rubin RJ, Schentag JJ, et al. The burden of Staphylococcus aureus infections on hospitals in the United States: an analysis of the 2000 and 2001 Nationwide Inpatient Sample Database. Arch Intern Med 2005;165:1756–61.
18. Wyllie DH, Crook DW, Peto TEA. Mortality after Staphylococcus aureus bacteraemia in two hospitals in Oxfordshire, 1997–2003: cohort study. BMJ 2006;333:281
19. Melzer M, Eykyn SJ, Gransden WR, Chinn W. Is methicillin-resistant Staphylococcus aureus more virulent than methicillin-sensitive S. Aureus? A comparative cohort study of British patients with nosocomial infection and bacteremia. Clin Infect Diseases 2003;37:1453–60.
20. Schrelonka RL, Infante AJ. Neonatal immunology. Semin Perinatol 1998;22(1):2–14.
21. Healy CM, Palazzi DL, Edwards MS, et al. Features of invasive staphylococcal disease in neonates. Pediatrics 2004;114:953–61.
22. Wiswell TE, Curtis J, Dobek AS, Zierdt CH. Staphylococcus aureus colonization in relation to device used. J Pediatr 1991;119(2):302–4.
23. Wiswell TE, Miller GM, Gelston Jr HM, et al. Effect of circumcision status on periurethral bacterial flora during the first year of life. J Pediatr 1988;113(3):442–6.
24. Fortunov RM, Hulter KG, Hammerman WA, et al. Community-acquired Staphylococcus aureus infections in term and near-term previously healthy neonates. Pediatrics 2006;118(3):874–81.
25. Thompson DJ, Gezon HM, Rogers KD, et al. Excess risk of staphylococcus infection and disease in newborn males. Am J Epidemiol 1965;84(2):314–28.
26. Bratu S, Eramo A, Kopec R, et al. Community-associated methicillin-resistant Staphylococcus aureus in hospital nursery and maternity units. Emerg Infect Dis 2005;11(6). Available from http://www.cdc.gov/ncidod/EID/vol11no06/04–0885.htm.
27. Sauer LW. Fatal staphylococcus bronchopneumonia following ritual circumcision. Am J Obstetr Gynecol 1943;46:583.
28. Kirkpatrick BV, Eitzman DV. Neonatal septicemia after circumcision. Clinical Pediatrics 1971;13(9):767–8.
29. Annunziato D, Goldblum LM. Staphylococcal scalded skin syndrome. A complication of circumcision. Am J Dis Child 1978;132(12):1187–8.
30. Woodside JR. Necrotizing fasciitis after neonatal circumcision. Am J Dis Child 1980; 134(3):301–2.

31. Bliss Jr DP, Healey PJ, Waldhausen JHT. Necrotizing fasciitis after Plastibell circumcision. J Pediatr 1997;31:459–62.
32. Enzenauer RW, Dotson C, Leonard T, et al. Increased incidence of neonatal staphylococcal pyoderma in males. Mil Med 1984;149(7):408.
33. Curran JP, Al-Salihi FL. Neonatal staphylococcal scalded skin syndrome: massive outbreak due to an unusual phage type. Pediatrics 1980;66(2):285–90.
34. Stranko J, Ryan ME, Bowman AN. Impetigo in newborn male infants associated with a plastic bell circumcision. Pediatr Infect Dis 1986;5(5):597–9.
35. Enzenauer RW, Dotson CR, Leonard T, et al. Male predominance in persistent staphylococcal colonization and infection of the newborn. Hawaii Med J 1985;44(9):389–90, 392, 394–6.
36. Zafar AB, Butler RC, Reese DJ, et al. Use of 0.3% triclosan (Bacti-Stat) to eradicate an outbreak of methicillin-resistant Staphylococcus aureus in a neonatal nursery. Am J Infect Control 1995;23(3):200–8.
37. Saiman L, Cronquist A, Wu F, et al. An outbreak of methicillin-resistant Staphylococcus aureus in a neonatal intensive care unit. Infect Control Hosp Epidemiol 2003;24:317–21.
38. Nambiar S, Herwaldt LA, Singh N. Outbreak of invasive disease caused by methicillin-resistant Staphylococcus aureus in neonates and prevalence in the neonatal intensive care unit. Pediatr Crit Care Med 2003;4(2):220–6.
39. Davies EA, Emmerson AM, Hogg GM, et al. An outbreak of infection with a methicillin-resistant Staphylococcus aureus in a special care baby unit: value of topical mupirocin and traditional methods of infection control. J Hosp Infect 1987;10:120–8.
40. Reboli AC, John Jr AF, Levkof AH. Epidemic methicillin-gentamicin-resistant Staphylococcus aureus in a neonatal intensive care unit. Am J Dis Child 1989;143(1):34–9.
41. Storms M. Circumcision and endocarditis prophylaxis. Arch Pediatr Adolesc Med 1998;152:138.
42. Committee on Infectious Diseases and Committee on Practice and Ambulatory Medicine. Infection control in physicians' offices. Pediatrics 2000;105:1361–9.
43. Grayson ML. The treatment triangle for Staphylococcal infections. New Engl J Med 2006; 355(7):724–7.
44. Nicolle L. Community acquired MRSA: a practitioner's guide. Can Med Assoc J 2006;175(2): 145–6.
45. Kaplan SL. Implications of methicillin-resistant Staphylococcus aureus as a community-acquired pathogen in pediatric patients. Infect Dis N Am 2005;19:747–57.
46. Task Force on Circumcision. Circumcision policy statement. Pediatrics 1999;103(3):686–93.
47. Isaacs D. Personal communication.
48. Van Howe RS, Robson WLM. The possible role of circumcision in newborn outbreaks of community-associated methicillin-resistant Staphylococcal aureus. Clin Pediatr 2007;46(4): 356–8.
49. Nguyen DM, Bancroft E, Mascola L, et al. Risk factors for neonatal methicillin-resistant Staphylococcus aureus infection in a well-infant nursery. Infect Control Hosp Epidemiol 2007;28(4):406–11.

Fitting In and Getting Off

Elective Adult Male Circumcision in the United States and Britain

Zachary Androus

Abstract The majority of male circumcisions in the United States are performed on infants upon parental request. Most scholarly attention towards American circumcision focuses on the neonatal practice; however, significant numbers of adult American men elect circumcision for themselves, and many men circumcised as infants elect further surgical adjustment of their penises as adults to meet their preference or standard for what they perceive to be a desirable penis. Illuminating insights into the American cultural values that underlie the persistence of non-therapeutic infant circumcision can be found in the expressions of desire for circumcision or re-circumcision offered by adult men who elect the surgery for themselves. Using primary ethnographic data, this paper surveys those justifications, which include the preference of sexual partners for circumcision (or re-circumcision); social conformity; aesthetic appeal of the circumcised penis; and sexual fetishization of the circumcised penis, the act of circumcision, or the experience of being circumcised.

A majority of men in the United States are circumcised, although it is not a large majority; the best estimate for a national average from the National Center for Health Statistics is 65%, although this number varies geographically.[1] This number only counts those individuals circumcised as infants. Thousands of men seek elective genital surgery every year, some for a first circumcision, and others for revision or adjustment of an earlier circumcision. This vague number is a conservative estimate, based on the limited available data about rates of adult circumcision. Between 1995 and 2000, one clinic in a southern US town of approximately 50,000 performed over one hundred adult circumcisions.[2] Reliable national statistics are not readily available, but from what little data is available, a highly speculative range from several hundred to a couple thousand men annually is not unreasonable.

This paper is the preliminary report of findings from an ongoing ethnographic survey of adult men in the United States who elect circumcision for themselves. This is a medical anthropological project, which is to say that there is a focus on

Department of Anthropology, American University, Lorenzo de Medici University; Florence, Italy

G. Denniston et al. (eds.), *Circumcision and Human Rights*,

the collection and analysis of qualitative data, and it is concerned generally with the interaction between the medical system and social and cultural values, with a particular focus on relations of power. There are several reasons why this is an important line of inquiry. First of all, the current medical and social scientific literature on adult circumcision is limited at best. Second, scholarly attention to male circumcision in Europe and North America tends to focus exclusively on infant circumcision, and while adult and infant circumcision are fundamentally different practices in certain respects, they are closely related in other respects. Third, and most importantly for medical anthropology, adult circumcision represents an intersection of social and cultural values and medical practice played out on the bodies of individual men. This intersection of embodied experience, bodily praxis, the individual, society, and medical practice bring into focus several of the primary concerns of medical social science in one practice.

In this paper, I do not directly address any of the issues raised by infant circumcision in the United States. In my view, the issues raised by infant circumcision are fundamentally about the individual right to bodily self-determination. I personally see no issues raised by the practice of adult elective circumcision that are not raised generally by any other adult, elective aesthetic surgery, and I see no direct connection to any of the issues that make infant circumcision a problematic practice from a rights perspective. As such, I want to make clear at the outset that my goals in this paper are to understand what motivates some men to undertake this particular surgery, and to give voice to their experience; my goal is not to challenge their justifications or anything about the practice itself, and I am not passing judgment on their choices. I am seeking to understand what is important to the people making the choices, and how that motivates them to make the choices they do. I am not opposed to circumcision in and of itself, and I take no issue personally or professionally with elective body modifications of any kind. From my way of thinking, the problems with infant circumcision have to do with rights and consent rather than anything about circumcision itself. So, I have no issue with men who elect this procedure as adults. And I feel a responsibility to the men who volunteered to participate in this research, a responsibility to make sure that their voices come through in my reporting in such a way that accurately represents their motivations and experiences.

When we talk about adult circumcision, we are talking about things that actual living people in the world have done and we are talking about experiences that other people have undergone. I see two fundamentally different ways of talking about elective circumcision: we can talk about it in terms that the men whose lives we are discussing would recognize and relate to, or we can talk about it in terms that the men themselves might not recognize as describing their experience. I take the first approach, and as such, my analysis is always grounded in the words of the men themselves. The most important window through which we have to look is provided by the stories that the men themselves tell. This is part of what distinguishes an anthropological approach. With all of this in mind, you will not ever hear me describe these men as deviant or bearing a false consciousness about the nature of circumcision or its effects, although their choices may indeed be transgressive by certain social standards.

I want to quickly talk about my sample and my methodology, and then I'll outline some of my preliminary findings and discuss their relevance. I have collected a terrific amount of data, and this paper represents just the first preliminary report of my findings because the data collection and analysis is ongoing. For this research, I focus exclusively on non-religious circumcision because Americans have their own social tradition of circumcision, and I am not convinced of the utility of including unrelated practices together in an analysis of meaning and cultural value; while they all have a modification of the genitals in common, they are motivated by different ideas about what is appropriate and why.

Although I have had over forty volunteers thus far, the data for this paper was drawn from a group of fifteen respondents. My data was collected from live ethnographic interviews conducted in person, or over the telephone, or using an Internet-based instant messaging program. In every case, the interviews were recorded or transcribed. Participants were solicited from an Internet discussion group dedicated to adult circumcision and also by referral. Informed consent was obtained from all participants, in accordance with the research design approved by the Institutional Review Board of American University. My survey is by no means exhaustive. My survey does not aim to be statistically representative, in part because it is impossible to know with any degree of certainty the total population of men electing circumcision. With a purposive sample this size, any statistical correlations between demographic variables like age or income and why an individual chose circumcision would not only be restricted to the sample itself, it could potentially distract attention from the real significance of this project, which is the focus on the meanings and motivations of the men themselves.

Accordingly, my methodology is based on qualitative data collection and analysis, using grounded theory and established techniques of narrative analysis. While I do collect some quantitative information, I largely eschew quantitative data analysis in favor of narrative analysis. Why do I shun statistical modeling of my data? Because modeling the mathematical relationships between various aspects of my participants' behavior and their identity doesn't really tell us much about the actual lived experience of any given individual in that group. Let me use an example from another study to make this point. Collins, et al.,[3] conducted pre- and post-operative surveys of men being circumcised as adults and reported their findings in the *Journal of Urology* in 2002. They found no statistically significant difference in pre- and post-operative sexual function in their sample of fifteen men. But what does that really tell us? Let's look closer at their data. Participant one in the Collins study reported the same level of sexual satisfaction both before and after his circumcision: in both cases he chose "neutral or mixed" level of overall satisfaction from the four choices available. Participant ten also reported a "neutral or mixed" level of satisfaction both before and after his circumcision. So far, so good. No change in level of sexual satisfaction reported from these two participants, this is consistent with the overall findings. If we look at the other responses of these two participants, however, something interesting emerges: participant one reported lower levels of sexual satisfaction for every specific item on the survey. Participant ten reported the same or higher levels of satisfaction for every specific item on the survey. So here we have two men, both circumcised as adults. For neither man did circumcision appear to change their overall level of satisfaction, but one man clearly has experienced negative sexual effects and another man clearly experienced positive sexual effects. If everything about sex is worse for someone, what good is it to lose that detail in the overall big statistical picture? Not much, in my opinion. Likewise, if everything about sex is better for someone, how well is that reflected in a finding of no statistical significance? Using mathematics to try and extrapolate general conclusions from a series of highly individual, subjective experiences is, in my opinion, a dubious undertaking when it comes to things like sexual satisfaction and bodily self-image, hence my insistence on qualitative data. Last point on methodology, the questions I use in my interviews and surveys are all open ended; no multiple choice, no scales of one to five. The few surveys reported in the medical literature have relied primarily on multiple-choice questionnaires.

The history of American circumcision's transformation from a nineteenth-century medical treatment to a twentieth and twenty-first century social practice is well known to most of you. While the justifications are no longer exclusively medical, the procedure itself remains medicalized, insofar as it performed by medical professionals, excepting those cases when a particular religious practitioner is required to fulfill religious requirements. A similar state of affairs is current in most cases of adult circumcision, in so far as doctors primarily perform the procedure, even when there is no clear condition to be treated. This moves adult circumcision into the realm of other elective, aesthetic surgeries, which are commonly performed in the United States.

Male circumcision is widely considered to be a social norm in the United States,[4] however, the complex composition of US society calls into question the validity of applying a concept like social norm theory to a population as large and heterogeneous as that of the United States. Even within an apparently homogenous group, such as those men electing circumcision for themselves as adults, there is a good deal of variability in both the characteristics of the men themselves and in their motivation for and experience of being circumcised as adults. The title of my paper, "Fitting In and Getting Off," refers to two major themes that appear throughout the narratives of my informants. These two categories are not mutually exclusive, of course, and it appears that, very often, fitting in is an important part of getting off. That is to say, that social conformity and the self-confidence and comfort that come from perceiving oneself as normal or ordinary is, for many men, an important component of successful sexual relationships. We can think of these two broad themes as existing together on a continuum; on one end of the continuum are men for whom getting circumcised is primarily a social conformity thing and on the other end of the continuum are men for whom it is related primarily to their sexuality. It is important to note that, based on my analysis, no one appears to occupy either extreme of this continuum exclusively; that is to say, adult circumcision never appears completely dissociated from either sexuality or social relations, but always incorporates aspects of each. And these two factors co-exist with several others in many of the men.

Individual experiences vary quite widely. For example, among my respondents were several British men. Two of these men, let's call them Peter and Greg, are within a year of age, both are in their early fifties at the time I communicated with them. Both have similar levels of education and are married, straight men. Both their fathers were circumcised, but when they were each born, circumcision was not available through the British National Health Service. Peter started thinking about circumcision when he was in his mid-twenties, and at age 49 decided for sure he would do it, having the operation at age 50. Greg, on the other hand, never thought about it until he was diagnosed with balanitis xerotica obliterans (BXO), following a year of painful and troublesome symptoms that responded poorly to conservative treatments. Peter wanted to be circumcised because he believed it would improve his appearance, and he reports now a delight in being naked and an improved self-image. Greg, on the other hand, reports feeling self-conscious about being different from most of his friends, and a corresponding avoidance of locker room type situations. Greg even feels a "little bit more naked" when he's walking around his own bedroom. Peter, on the other hand, now happily attends nude beaches to show off his exposed glans. Both men report improved sexual relations with their wives following the procedure.

The history and current condition of routine infant circumcision in Britain is very different from that of the United States, but these two gentlemen illustrate an important point, namely that it is impossible to identify particular variables that will predict whether a man is or is not inclined to seek circumcision as an adult. Similar demographics, but Peter wanted circumcision because he liked it, while Greg would not have considered it had it not been for a medical condition. And, it isn't really possible to identify variables that might predict, with any kind of certainty, how a man will feel about the procedure once it's done; Greg felt that he had no choice in the matter because of his condition, but he reports no regrets and some definite advantages, especially in terms of his sexuality.

When US parents are making the decision whether or not to have their newborn sons circumcised, the circumcision status of the father appears to play an important role in their determination. However, the condition of the father is not determinative, which is to say that some uncircumcised men have their sons circumcised and vice versa. When I first undertook this research, one of the things I was most interested in exploring was whether the circumcision status of a man's father correlated in any way with his decision to seek circumcision for himself. While my US-born informants were somewhat more likely to have been left intact if their father was intact, they all share an appreciation for the aesthetics of the circumcised penis, and a desire to fit in and appear normal, as well as acknowledgment of the erotic and sexual appeal of being circumcised.

One of my informants named Bruce wrote, "I truly wish it had been done at birth." He was born in the rural Midwest in the early 1950s, and neither he nor his father had been circumcised at birth. Circumcision, according to him, was uncommon in his region at the time he was born. However, he also reports being embarrassed at the sight of other boys when he was growing up, so, while it may not have been common, it certainly was not unheard of by that time. Bruce reports that he first started thinking about getting circumcised when he was a teenager, and that he was too embarrassed to even try and have sex with anyone until he was himself circumcised in his early thirties.

Regret for the decision is not expressed by any of the men who volunteered to participate, but there are mixed or ambivalent responses from some. This does not always have to do with the decision itself. Cliff, for example, is an educated gay professional in his mid-60s who was born and lives in the Pacific Northwest. He was circumcised as a child, but with very little skin removed, leaving him with an unsatisfactory appearance. At the age of 20, he began considering circumcision, or re-circumcision if you will, and finally carried it out at age 64. After doing quite a bit of research and discussing the procedure with others online, Cliff requested his urologist to perform a "high and tight" circumcision that left his frenulum and as much inner foreskin as possible intact. This request was based on Cliff's experience of the inner foreskin as "the home of sensitivity" and his discussions with other men circumcised as adults. But Cliff was disappointed with the results because the urologist removed a good deal of inner skin, although he still reports an improved self-image.

It appears to me that Cliff found himself caught in the uncomfortable intersection between medicine and culture at which circumcision dwells in the United States. Cliff says his main reason for wanting an adult circumcision was the visual appearance. A urologist is not an aesthetic surgeon though, and throughout the discussion group in which I met Cliff, men described urologists as not being as concerned with the visual outcome. However, the visual outcome is clearly very important to many people. This situation raises issues that are commonly discussed in the context of other aesthetic surgeries, such as breast augmentation or rhinoplasty, namely the tension between psychologically therapeutic surgeries and physiologically therapeutic surgeries. If a surgical modification that is not physiologically necessary carries profound mental and emotional benefits, is it medically therapeutic? Is it appropriate for doctors to perform?

There are no simple answers to these questions, but they are raised by the experience of people like Cliff, and they are important questions for both medicine and social science.

Cliff's experience also represents the continuum of motivating factors that I mentioned earlier. Cliff says the primary reason he wanted to be re-circumcised was visual; but he also says that he finds circumcision itself to be erotic. He wanted to watch the procedure as it was being performed, but his urologist would not permit this. Instead, Cliff was sedated and put in surgical restraints, which, according to him, "was just not the way I wanted it to happen." Cliff's desire to be awake and to see the procedure represents a challenge to the traditional authority of doctors over the material details of a surgical procedure. Cliff wanted to watch because, for him, getting circumcised was about more than just the results of the procedure, it was also about the experience of the procedure itself. This tension mirrors the conflict I just discussed between elective aesthetic surgery and surgeries deemed physiologically necessary by physicians. Who decides what kinds of procedures are appropriate, and who decides what are appropriate ways to carry out those procedures? The answers to these questions are constantly negotiated between individual patients and doctors. I can give you some overview of the trends in my sample.

Eight of the fifteen men reported that aesthetics or appearance were their primary reason, two reported social conformity as a primary reason, and only one reported sexuality as a primary reason. Six men reported social conformity as a secondary reason, and four men reported something related to sexuality as a secondary reason. No one who reported social conformity as a primary reason gave aesthetics as a secondary reason. Rather than crunch these numbers to model findings statistically, I think it's more useful to frame the results in terms of the continuum I mentioned between social conformity and sexuality, between fitting in and getting off.

References

1. National Center for Health Statistics. 2005. Trends in circumcisions among newborns. Electronic document, http://www.cdc.gov/nchs/products/pubs/pubd/hestats/circumcisions/circumcisions.htm, accessed July 28, 2006.
2. Fink KS, CC Carson, and RF DeVellis. 2002. Adult Circumcision Outcomes Study: Effect on Erectile Function, Penile Sensitivity, Sexual Activity and Satisfaction. Journal of Urology 167(5): 2113–6.
3. Collins S, J Upshaw, S Rutchik, C Ohannessian, J Ortenberg, and P Albertsen. 2002. Effects of Circumcision on Male Sexual Function: Debunking a Myth? Journal of Urology 167(5): 2111–2.
4. Waldek S. 2003. Using Male Circumcision to Understand Social Norms as Multipliers. Cincinnati Law Review 72(3): 455–526.

NORM-UK

David Smith

Abstract A positive or negative image can have a huge impact and greatly influence how people think. In a fast-moving, short-attention-span world, first impressions play a major role. Should we be advocating the phraseology of "anti-circumcision" or "pro-foreskin"? Both are equally valid but create a vastly different impression in the mind of the public. We explore the ethics and ethos of this question from the British perspective through the foundation and development of NORM-UK.

"There is hardly a reason to circumcise a little boy for medical reasons because those medical reasons don't exist," Dr. Michael Wilks, Head of Ethics at the British Medical Association, said on a BBC World Service program, who admitted that doctors have circumcised boys for "no good reason." He also said that the majority of people who have been circumcised in the past, for what were put to them or their parents as good medical reasons, probably were no such thing, and those people certainly have a right to make a claim that what was done to them was an unnecessary and premature intervention at a time when they had no capacity to object or no say in the matter.

NORM-UK was founded in 1994, as a direct response to a significant need in Britain for an organization concerned with circumcision and the detrimental effects occurring from it. The aspiration was to raise awareness about the various problems that arise from this invasive procedure. Up to this point, circumcision had not been a topic of conversation in the United Kingdom, and any man affected by it had to suffer in silence. NORM-UK's activities were expanded into foreskin restoration, counseling groups, and outreach work based around the subject of circumcision.

Just as circumcision had not been widely discussed, it soon became apparent that neither had the foreskin. NORM-UK was bombarded with requests for information by intact men with a foreskin problem, who knew that a visit to their doctors inevitably would result in a circumcision they did not want. As a direct result, NORM-UK has become a multi-functional organization dealing with all aspects of foreskin health, alternatives to circumcision, and foreskin restoration.

General Manager, NORM-UK, Stone, Staffordshire, UK

G. Denniston et al. (eds.), *Circumcision and Human Rights*,

Organizations change over time and, since becoming General Manager, I consider my task is to move our organization forward. As part of this exercise, I have had to look afresh at every aspect of our activities. Part of this is to look at how circumcision and the foreskin are perceived in the United Kingdom as compared with other nations, in other words, review our image.

Unlike the United States, circumcision is fortunately now not widely practiced in the United Kingdom. Since the formation of the National Health Service in 1948, statistics show that the circumcision rate has gradually dropped. According to a report in the *British Journal of Urology* in July 2006, circumcision rates in England continued to fall up until 2000, particularly in those under five years, in whom pathological phimosis is rare. However, it is still being performed far more than is necessary because medical professionals are not taught about alternative treatments. The vast majority of the British public currently think that circumcision is something done by the Jews or for genuine medical reasons.

The biggest problem facing NORM-UK is the ignorance, both of the medical profession about alternatives and the public, both about the damage caused by circumcision and, of more significance, the importance of the foreskin and the lack of information on simple alternative treatments.

The British Association of Pediatric Surgeons (BAPS) recently convened a working party on the Management of Foreskin Conditions. NORM-UK and Doctors Opposing Circumcision (DOC) have been invited to contribute by submitting documentation, but our requests for a representative on the working party have been refused. Unfortunately, it is still the case that representatives of people undergoing surgical procedures are not allowed input into discussions about their treatment. At the moment, the working party has produced a draft document of more than 40 pages, on which we have been invited to comment. The British Medical Association has made their position very clear, stating that the medical benefits previously claimed have not been convincingly proven. Now, it is widely accepted, including by the BMA, that this surgical procedure has medical and psychological risks, which is a small step in the right direction.

In the past, the *Government Response to the Reports of the Health Committee on Health Services for Children and Young People* stated: "Surgical interventions should only be performed when clinically necessary, especially in children. Yet, in April 2006, we received a copy of a report stating that children at the Royal Hospital for Sick Children in Yorkhill, Scotland, are facing longer waits for operations because of a massive backlog of religious circumcisions. A spokesman for the pediatric surgeons at Yorkhill said they did not want to encourage anyone else to do the operation because they see a handful of cases every year where the circumcision has been done outside the hospital and the child turns up with an infection, bleeding, or an unsatisfactory result and that these are people whose lives have been ruined by having a badly done circumcision in childhood."

It is obvious that, no matter how many guidelines, reports, and recommendations are produced, the hospitals and doctors are going to continue to ignore them. Part of the problem is that these people seriously think it is only procedures conducted outside the hospitals that cause problems and ruin lives, not, as we are aware, that all circumcisions have a destructive and adverse effect. They are continuing to deny the damage caused by circumcision because they do not know enough about the function of the foreskin.

In Britain, we have lived with the pro-circumcision brigade handing out their propaganda message for too long. Circumcision has always been the cure; it is only the disease that has changed. In the past, we have had the tendency to concentrate on condemning circumcision. This route has disadvantages. In making the case against circumcision, the majority of the time has been spent discussing the negative aspects of this operation.

- We use the "C" word a lot, which can be a trigger for some.
- We get involved in the religious debate, whether we like it or not.
- Debating the "supposed" benefits of this operation with the medical profession is reactive.

Only a small amount of time has been spent informing people about the foreskin, its functions and beneficial qualities, which is neglecting the most important aspect in this debate — the foreskin.

Public opinion is a very powerful tool that can be used to our advantage. Public opinion can force people in power to change the way things are done. Knowledge is power. The public in general is ignorant about the foreskin, why it is there, how to take care of it, how it works, and what it provides, both for the child and for the man. They are also ignorant about what is lost when the foreskin is amputated or the problems that can occur in later life. If we educate the public, they then are able to make the informed choice that is supposed to be offered by the medical profession.

If we cannot get the medical profession to stop recommending circumcision, we need to educate parents about the foreskin so that, with knowledge, they refuse it. Inform men when they are teenagers about the importance of their foreskin, so they value it more. Most men are told that their foreskin is a useless piece of skin and, because they are told by people, in general doctors "who know about these things," they believe them.

Challenging the medical profession and religion can be like banging your head against a brick wall. We have a lot of work to do through education, providing easily accessible, accurate information through the media, and drawing attention to the subject universally.

NORM-UK's aim for the future, especially in Great Britain, is to ensure that "Joe Average" will conclude for himself that, if foreskin equals good, then circumcision equals bad. This approach can also hold its own in the religious debate because, in the end, all we are doing is promoting the positive aspect of the foreskin. Whilst not mentioning circumcision at all, everyone is thinking about it! If challenged, we just reiterate the proven scientific facts about the functions of the foreskin. This way, no organization or individual can be accused of being anti-religious.

A campaigning organization, especially in the United Kingdom, has the significant problem of limited opportunities for funding. NORM-UK is classed as an educational charity, which has the advantage that some money is available for suitable projects.

Since the formation of NORM-UK, we have actively taken pro-circumcision doctors to task, pointing out that they are not complying with current guidelines. Often, our letters are answered with abuse, such as the doctor from Surrey, who responded to a follow-up letter by saying that he did not reply to our original letter because it was clearly the work of "a deranged mind." This particular doctor was running a private circumcision service for infants, charging about $500 per circumcision. Others have been more open in their attack. The *BMA News Review* published a "debate" between Dr. Janet Menage and Dr. Nigel Zoltie. Zoltie is credited only with being an accident and emergency consultant in Leeds but, in truth, he is also a ritual Jewish circumciser, as well as a member of the Initiation Society. He revealed this crucial fact in a published letter to the editor of the *BMJ*. This presents an obvious conflict of interest. Since Zoltie has a pre-existing bias in favor of circumcision, complicated by a presumed financial incentive in promoting circumcision, his "medical" opinion on circumcision is difficult to distinguish from a marketing campaign. Zoltie writes: "Those who criticize the actual operation use emotive words like mutilation, and ascribe a wide variety of complications to the surgery. However, mutilation is in the eye of the beholder. What is mutilation to one observer may be beautiful to another, pierced body parts, for example."

These doctors are allowing their religious beliefs, avarice, or money-making activities to take precedence over their Hippocratic Oath. *The Declaration of Geneva* requires a doctor to put the health of the patient as the first consideration, and not to allow race, religion, or nationality, to intervene between his duty and his patient, nor shall a doctor use his "medical knowledge contrary to the laws of humanity."

We sometimes think the situation is bad in the United Kingdom, but in the United States, the move against circumcision has been met with more hostility by the medical profession.

NORM-UK was founded over ten years ago and, when it was founded in 1994, circumcision was the ultimate taboo. It was difficult to get any sort of publicity on the subject. The breakthrough came in December 1994, when a mention of NORM-UK was made on a national radio program. In those days, any program or article featuring details of men unhappy about circumcision usually included an interview either with a doctor or rabbi, ridiculing opposition to circumcision. This piece included an interview with a urologist from Stepping Hill Hospital in Cheshire, in which he stated he was afraid that, if foreskin restoration caught on, it would "pander to a lot of potentially inadequate people." This did not stop these "potentially inadequate people" contacting NORM-UK. Even now, ten years later, members of the medical profession still occasionally make this sort of crass remark but, pleasingly, the situation with publicity has changed over the last decade. Although there have been a number of very positive articles in the British press, the media still seems reluctant to tackle the subject, which is something we are actively addressing and determined to overcome.

Our approach is on several fronts, including medical campaigns, education to both the young and adult alike, dissemination of information to the medical profession, midwives, and health visitors, and most especially ensuring that the name

NORM-UK is instantly recognizable and associated in every male's mind, not only with the provision of information but as someone from whom they can obtain, whenever needed, accurate, up-to-date medical advice and support.

In looking at our image in the twenty-first century, we need to embrace modern technology. The World Wide Web has revolutionized communications. Used properly, it can be our best public relations resource. It enables a national organization to become a worldwide organization, as the name implies. Although the bulk of enquiries to NORM-UK are from the United Kingdom, we are now receiving an increasing number of emails from other countries, India in particular. Electronic communication has the advantage that, not only is it instant, but it is also anonymous. It is possible, for example, for a concerned teenager to be referred to a doctor without disclosing his identity. The downside to the World Wide Web, from a health information point of view, is that we are competing against a sea of data from often-unreliable sources. Someone searching for help to enable them to solve a foreskin problem may have difficulty in ascertaining who is giving genuine approved advice, sifting the fetish from the genuine medical. The problem is compounded further by the fact that even genuine medical sites are giving out-of-date information. We also have to contend with the fact that resources may not be specific to the country of enquiry. A teenager from the United Kingdom, seeking details about circumcision or foreskin health, is quite likely to encounter an American webpage, where he would be given very different advice from a United Kingdom-based site.

Our present website has served us well, but was designed a few years ago and we plan a total revamp, possibly creating a number of different sites.

We need to place a greater emphasis on ways to improve our online credibility. There are a few strategies we can employ to do this. The first is to ensure that our site displays as many accreditation schemes, awards, and logos as possible. The HON (Health on the Net) and CHIQ (Centre for Health Information Quality), for example, and other general design/content awards and logos from respected bodies will add to our respectability.

With regard to site design and dynamism, we need to have a professional and up-to-date look. Teenagers particularly are web-savvy; an amateurish site will infer poor information. Dynamics, such as recent news updates, improves the immediacy and relevance of the site in the eyes of the visitor, as does randomizing some elements of content on the front page.

With regard to search engine placement and appearance, we consider it is worth spending some time with search engine optimization programs and doing some research on the topic. Where we appear on the search results and how we appear are important. The excerpts shown on results pages are taken directly from the sites, and it is not difficult to influence them. The best will appear immediately relevant to the searcher. With this strategy, we don't need to be at the top of the results, but should ideally be in the top half of the first page. Also, we need to try to aim specific pages at specific search keywords and make them appear relevant. We need to remember that our audience is searching for the best information — we need to convince them that we have it.

We also need to ensure that we have as many links as possible from trusted sites. Links from almost anywhere are good for search engine placements and visitors, but those from trustworthy health organizations, such as the Men's Health Forum in the United Kingdom, are especially good for credibility. It's also worth emailing any website that links to "bad" sites, as it probably won't take much for them to take the links down (or even replace them with ours).

From the emails, letters, phone calls, and requests for information, it is evident that one of NORM-UK's strongest qualities is its role in advocacy. This can range from direct dialogue with the powers that be, the Chief Medical Officer, the Men's Health Forum and supportive Members of Parliament to adjoining the organization that provides sex education for teenagers in British schools.

Although NORM-UK is currently run from a single office in the center of England, its impact and information dissemination is global. NORM-UK is totally independent from its American counterpoint, but obviously there is considerable empathy and cooperation between these organizations on both sides of the pond. NORM-UK has developed a network of support resources, readily accessible fact sheets, and, above all, instant medical assessment and confidential advice, which has helped many who have turned to us at the eleventh hour. Whereas our ambition obviously is to eradicate the procedure, the more immediate goal must be to ensure that the organization's name is as universally recognized for its activities as Asda/Walmart or the British Broadcasting Corporation.

During the past decade, dedicated trustees and volunteers who believe in the philosophies and strategies of NORM-UK, have gathered arguably one of the finest accumulations of data (made totally anonymous) that we hope to harness in the future. Similarly, the organization prides itself on its archive resources and ability to supply answers, not only to the enquiring public but particularly to the ever-growing requests from the media. Barriers steadily are being broken down by those who believe absolutely in our subject but then have bravely, as it were, put their head over the parapet and spoken openly to a national audience on the wide-ranging issues surrounding circumcision and its aftermath. This is forcing open the door that previously had been closed firmly to our subject, particularly as a result of the enduring victim attitudes still socially prevalent in the mindset of the majority of the public, to whom anything related to sex was spoken about in whispers behind closed doors.

In 1806, we still had slavery, but only just. By 1856, it was banned. In 1906, we would have been agreeable to no votes for women. By 1956, we had moved on. In 2006, sadly there are many who accept genital mutilation, even without anesthetic. Let us hope that considerably sooner than 2056, things will have changed for the better.

Real Men: Foreskin Cutting and Male Identity in the Philippines[1]

Leonard B. Glick

Abstract Most Filipino boys submit to foreskin cutting as an essential experience in the transition from childhood to adulthood. Usually this means not circumcision but supercision, which consists of a single dorsal incision with relatively minimal tissue destruction. As a further important contrast to the situation in the United States and other Anglophone countries, in the Philippines foreskin cutting is seldom touted for its ostensible medical benefits; rather, the practice is embedded in broadly accepted social norms connected with male identity, social maturity, and sexual acceptability. Moreover, although beliefs about cleanliness are part of the picture, the dominant theme is not *foreskin rejection* but *penis improvement*, and the anticipated reward is not disease prevention but social acceptance as a properly formed man. There is good evidence that supercision is an indigenous practice that long antedated the arrival of either Christian or Muslim missionaries, and that although Islamic circumcision replaced the older practice, the Christian population retained supercision. Filipino beliefs about foreskin cutting correspond closely with those of Polynesians, who also practice supercision. In all these societies, foreskin cutting is so firmly embedded in entire cultural systems that it will probably endure until the cultures themselves change radically.

Introduction

In a short popular account of life in the Philippines, F. Landa Jocano, a pioneering Filipino anthropologist, recalled having submitted to foreskin cutting in his home town on the island of Panay:

> I was among a group of five boys one hot noonday, resting under a jackfruit tree, when Indo, the oldest among the boys, started teasing Osi, the youngest, that "it is about time you are circumcised." Osi answered that it was not necessary, for he had heard his uncle say that Malitbog girls prefer uncircumcised men. Indo went on: "Ah, I do not believe you. You are just afraid — like the rest of the boys." There was a chorus of protest. "All right, if you are not afraid, let us go to Itek and be circumcised," he challenged the group. There

Professor Emeritus of Anthropology, Hampshire College, Amherst, Massachusetts, USA

> were hesitations and protests. Osi finally said, "I will go if you will also be circumcised." The boys all agreed to go to Itek if Indo would submit to the operation. Afraid that he would be accused of being a coward himself, Indo said he would even be the first to undergo the cutting, "provided all of you follow afterwards." The rest agreed, and the five boys were circumcised that afternoon.[2]

Jocano described "circumcision" (probably supercision, see section below titled An Overview) as "the only rite which males undergo" during adolescence. This is unlike any "rite" an American boy might experience: "Boys may group together and decide to undergo the operation," he noted, "requesting a specialist for the purpose." They are often led to "submit" by being subjected to "teasing" and challenges to "their ability to withstand the pain of the operation."

Being cut was not in itself sufficient to avoid taunting; one had to be cut at the correct time — in Jocano's town, age eight to ten. Earlier supercision was said to be "bad for the health." And, as if that were not sanction enough, a "very young boy who is circumcised is made the object of the endless fun and ridicule of his playmates." One boy, circumcised or supercised at birth in a hospital, was "very unhappy." Boys his age would shout at him, "Here comes the circumcised, the circumcised." He is said to have avoided boys his own age and to have played either alone or with younger boys and girls.

Aside from the obvious social pressure to accept foreskin cutting, Jocano's brief account hints at two definitive elements in Filipino beliefs and attitudes to the procedure. First, boys had to be cut "at the correct time" — i.e., when they were at or near puberty. Those cut at too young an age were violating established social boundaries between childhood and sexual maturity. Second, although the uncle's comment, that local girls preferred intact men, runs counter to other statements about Filipino beliefs, it does reveal that a central consideration here is potential sexual desirability. Social maturation, future success as a sexual partner: these are foremost in mind when a Filipino boy agrees to have his foreskin cut.

For those few boys who remained intact, says Jocano, ridicule led to "deep-seated fear of circumcision." He cites the case of a 14-year-old boy who went to work at a reforestation nursery. One day he went with other workers to bathe in a river; and when the other men saw that he was intact they "teased him very roughly about it." The game continued, with the men declaring that they themselves should circumcise the boy, but if he resisted, "they would have him hog-tied." After a few days of such torment, just after one man proposed circumcising the boy while he slept, he left for home. Asked later by the nursery director why he had left, he replied, "Because the men are determined to circumcise me if I remain there."[3]

An Overview

Jocano's experience was typical for Filipino boys. Although most accounts use the term "circumcision" for all cutting, supercision — cutting a longitudinal slit on the dorsal (upper) side of the foreskin and folding the tissue down — is the most common operation.[4] This is significantly less radical than the American practice; foreskin *cutting* in the Philippines usually means relatively little foreskin *removal*.[5]

Information on this subject has been spotty; and even now one must depend on scattered ethnographic sources, occasional newspaper articles, and the work of a few pioneering researchers — a reflection of the fact that most Filipinos take foreskin cutting for granted. Jocano's account and others show that most boys accept the experience as a painful — though necessary and tolerable — ordeal. Nevertheless, Filipinos differ significantly from Americans in their beliefs not only about how foreskins should be cut but even about what the cutting accomplishes. Filipino beliefs about foreskin cutting are intimately connected with their beliefs about maleness and male sexuality.[6] A Filipino boy must be cut to become a "real man" (*totoong lalaki*)[7] who will be acceptable to women as a lover or husband. Pressure on boys, not only from adults but also from their peers, is so intense that refusing to conform condemns them to ostracism, ridicule, and — perhaps most significant — rejection as potential lovers or husbands. Foreskin cutting is so embedded in broadly accepted social norms that approaching it as though it were detachable from its cultural meaning would be seriously misleading.

Geography, Demography, and History

The Philippines are composed of many islands, some large or medium-sized, others very small. The main island in the north is Luzon, which contains the capital city, Manila; just to the south of Luzon is the large mountainous island of Mindoro. In the center are the Visayan islands, including Cebu, Leyte, Negros, Panay, Bohol, Samar, and the outlying Palawan. At the southern end are the large island of Mindanao and the numerous smaller islands of the Sulu archipelago.

The people are physically diverse, products of centuries of ethnic mixing. The dominant physical type can be roughly described as "Malay," but many people (so-called "Mestizos") have Chinese and/or European ancestors as well. More noteworthy is the country's social, ethnic, and religious diversity. The population is now approximately 83 million; about half live in cities; the rest are mostly peasants. Some 80% of the population are Catholics, centered in Luzon and the Visayans but nowadays migrating in increasing numbers into Mindanao; another 9% are Protestants. But although almost 90% of Filipinos are nominal Christians, elements of traditional ("folk") religion, including rituals to supplicate or appease deities and spirits of every description, survive in many communities as part of daily reality.[8]

Muslims, nearly all of whom live in the southern provinces, constitute about 4% of the population; they practice versions of Islam also blended with beliefs and practices inherited from traditional religions. Non-Christian ("tribal") peoples make up the remaining approximately 7%. Most live in the hill country of northern Luzon, while a few inhabit the mountainous interiors of Mindoro and Mindanao. These people do not differ significantly in physical appearance from other Filipinos;

they appear to be historically related to coastal peoples and adhere to cultural practices that may have been dominant before the arrival of the Spanish. In recent decades, missionary activity and pressure toward "modernization" must inevitably have influenced the cultures of hill peoples.

Philippine political history is too complex for more than brief mention here, but its most salient feature — foreign invasion and occupation — may be relevant to our immediate subject. Humans have lived in the Philippines for at least 25,000 years. Commercial contacts and cultural interchange with Formosa, China, Malayan and Indonesian kingdoms, and India were a prominent feature of Philippine history for the past 1,500 years or more. The region had become a part of the Southeast Asian cultural area long before the arrival of Europeans, and it remains so today.[9]

Muslim missionaries from Indonesia reached the southern islands (Sulu Archipelago and Mindanao) by the fifteenth century, and the impact of their presence has endured there. Spanish colonization, concentrated in the central and northern islands, began in the sixteenth century. From then until the nineteenth century, there were repeated resistance movements, none formidable enough to defeat the occupiers. But by the final decade of the nineteenth century much of the population was turning to open revolt. Then, in 1898, in a momentous accompaniment to the Spanish-American War, Americans captured Manila, and soon afterward the Spanish ceded the country to the United States. From 1899 to 1902, Philippine insurgents fought a bitter war against the Americans, but they were brutally repressed and defeated. Political independence was for decades a contentious issue between the Philippines and the United States; only in 1946, following the severely disruptive events of the Second World War — Japanese occupation, American reconquest — was complete independence finally granted. American economic ties established during a half-century endure, as have some cultural influences — most notably, widespread use of English, particularly in urban areas. Nevertheless, the Philippines are an entirely distinctive nation, with historical memory, social customs, and cultural traditions wholly their own.[10]

Supercision in the Central and Southern Philippines (Visayas and Mindanao)

Although most accounts of Filipino genital cutting use the term "circumcision," the fact is that most boys are supercised. This is a distinctive practice, historically and geographically distinguishable from circumcision. It seems certain that supercision was the original practice in many indigenous Filipino societies. In an authoritative portrait of sixteenth-century Philippine culture and society, the late historian William Henry Scott reported that supercision was widely practiced in the Visayas before the arrival of Europeans:

> It was called *tuli* and was technically supercision rather than circumcision — that is, cut lengthwise above rather than cut around. … The uncircumcised were called *pisot*, an unripe fruit or green youth, a term which was also a polite euphemism for the female parts. The

operation was performed informally with no particular ceremony, and was thought to serve hygienic purposes.

Spanish missionaries were quick to conclude that the custom had been introduced by their Muslim competitors, as it probably had been in some areas, but Visayans claimed that their custom was of indigenous origin predating any contact with Islam. ... In the seventeenth century, however, the word *islam* came to be used in Visayan for circumcision according to the Muslim rite, and *magislam* meant to perform the ceremony.[11]

Why would Spanish missionaries have believed that Muslims, who advocated only circumcision, had introduced supercision? It seems much more likely that the original practice everywhere had been supercision, and that Muslim missionaries urged converts to practice circumcision instead.[12] Supercision was a widespread practice in Polynesia (although now replaced in some societies by circumcision) and is also found in a few other Southeast Asian island societies. It was probably the original form of foreskin cutting in much of the Pacific region.[13]

In a recent report on foreskin cutting among men from Cebu City (in the Visayas) and Davao (the urban center of Mindanao), Romeo Lee and Loyd Norella remarked that the most typical method is a "dorsal slit" (i.e., supercision), which one physician considered "the simplest." They diagram what are described as "the most commonly used dorsal slit" and "the unfamiliar *girlo* or coronal cut" (circumcision).[14] The report includes a description of supercision "in a community-based, non-clinic setting," accompanied by an observer's simple line drawings:

The circumciser gave the boy guava leaves and asked him [to] keep on chewing them. Then he told him to sit on a stool. The circumciser inserted the *cobra* [L-shaped piece of wood] through the dorsal side. Using a knife, the circumciser cut the prepuce lengthwise. His assistant pounded the knife with a grinding stone. The circumciser folded the prepuce back to expose the glans. He asked the boy to spit the chewed guava leaves and placed the leaves on the cut skin. The circumciser then dressed the cut penis.[15]

Note that, aside from the damage inflicted by cutting and "pounding," this procedure involved no removal of foreskin tissue.

In 1971, researchers conducting a "Family Health Project" in Cebu described how supercision was being performed then:

Between nine and eleven years of age, cohorts of Cebuano boys approach a supercisor (*manalit*) to request him to perform the relatively simple operation. The tools used are a sliver of bamboo, a large "bowie" knife, and either wooden mallet or large stone. The sliver is placed under the foreskin in longitudinal fashion so as to form a striking base. The foreskin is then stretched over the sliver, and the knife blade pressed hard on the outstretched skin so as to split the latter. An alternative method is to stretch the foreskin over the sliver, rest the blade on the skin, and strike the latter a sharp blow, with either mallet or rock. The wound is then covered with ground coconut powder, and the boys are advised to run to the sea and bathe.... The entire procedure is notable for its simplicity and lack of either religious or secular ritual.

In recent years, they note, parents have been increasingly taking their sons to local hospitals, "where this form of genital mutilation" is performed "under more hygienic conditions." Interviews with physicians in the city and its environs confirmed that hospitals and clinics were performing only supercision. They add that, although most urban people did not believe in "ritual purification" or "magical considerations," supercision was "supported by other powerful sentiments."[16]

F. Landa Jocano has described the culture of lowland village people in Panay and the adjacent island of Negros. Villagers believe that youths should not be cut until they are "well-advanced in adolescence." Here again, as Jocano reported for his own town on Panay (using some of the same phrasing), foreskin cutting in early childhood is considered "bad for the health," and young boys who are already cut are "made the object of endless fun and ridicule." This is a significant clue to Filipino ideas about the purpose of foreskin cutting. Boys are supposed to be supercised only when they have arrived at an appropriate time for entry into early manhood. *The event is not thought of as a medical procedure; it is an initiatory experience, a prerequisite for admission into adult status and sexual acceptance by women.*[17]

Although most residents of the southern island of Mindanao are Christian or Muslim, ethnographers have written about peoples living in the mountainous interior who still maintained their traditional way of life at the time of study. One such group are the Manuvu' (also known as Manobo or Bagobo), whose culture was studied by E. Arsenio Manuel in the 1950s and 1960s. The Manuvu' were originally semi-nomadic and nonliterate, but by the time of Manuel's study they had been heavily influenced by contact with Americans and other outsiders, and their culture had already changed substantially. Nevertheless, they remained ethnically distinct. At one time, tooth filing, tatooing, and supercision were all considered essential for entry into manhood. But even after the other practices had been abandoned as excessively "painful," supercision was retained because it required "a very much shorter period" and was a "less painful experience."

Manuel described the procedure and its rationale: The operation consisted of "stretching the prepuce over a stick provided with a pad of abaca fiber [manila hemp] to secure maximum cutting efficiency which is done with a sharp knife in one stroke." Nevertheless, Manuel adds, "there is that excruciating feeling." Why then do boys not resist? Because they "are told by the older folks that when they get married their wives would not like them with an uncut prepuce. The male sex organ becomes dirty and odorous, people say, and during intercourse the act would be scandalous, what with the lapping sound it would produce (like a lapping dog). Then he [i.e., the intact young man] is called *luop*, the uncut one, in the community; also, *tuppason*, the dirty one, because his organ is odorous. The female sex reacts in the same manner."[18]

Finally, in a brief memoir published in March 2005, a man in his forties recalled his childhood experience in Cotabato City, a rural town in Mindanao. The author, supercised in company with other boys, says that this still happens in his region. Here is his cheerful description of what he recalls, and of what appears to be contemporary practice:

> With eyes focused, an old man positions a "pinute" (knife) under my stretched foreskin while extending his other hand holding a wooden club. In the blink of an eye, the man delivers a quick blow. "Whack" goes the club as it hit the back of the blade and fresh blood dripped from between the legs of a boy as peers waited for their turn, anxiety painted on their face [sic]. The boy, smiling yet visibly shaken, quickly jumped into a nearby stream.

This author added a comment that I've encountered nowhere else. Note that, as usual, he identified his experience as "circumcision" even though he described supercision.

> In the Philippines, where it has been a tradition for over a century now, there is nothing definite as to how and when it really began. But common belief has it that the practice was introduced by western colonizers. There have been talks that colonizers used circumcision to identify groups supportive to them, even as history books appear to have failed mentioning it.[19]

Supercision in Luzon

There are few reports of supercision in Luzon, and I've seen none from Mindoro. Jocano's 1982 ethnography of an Ilocano community in northwestern Luzon includes a full description of supercision. The Ilocano are linguistically distinctive but part of the Christian population of Luzon.[20] The Apayao, or Isneg, a non-Christian people living in far northern Luzon, also practiced supercision on boys at "about age ten" and viewed intact men as "unclean and offensive to women."[21]

In northern Luzon live a number of semi-isolated ("tribal") peoples (Kalinga, Ilongot, Ifugao, and others) speaking distinct languages. Most are now significantly influenced by acculturative pressures, but some still maintain elements of their traditional cultures. Though much of the ethnographic information for these societies is now dated and the accounts are ambiguous or silent with regard to foreskin cutting, it appears that the practice may have been either uncommon or absent in their traditional cultures.[22]

One exception for northern Luzon is the ethnographic research conducted by Edward Dozier among northern Kalinga communities in 1959. He was told that they had never practiced foreskin cutting, but that among southern groups (living nearer to Christian populations) it had been customary in the past (i.e., earlier than the 1950s) for boys to be cut "at about the age of seven." But by the time of his sojourn, Dozier concluded that the operation was "rarely performed" by any Kalinga.[23]

Two ethnographers mention foreskin cutting among the Agta, or Aeta, also of northern Luzon. John M. Garvan, who studied these people (called "Negritos" in his work) in the first decades of the last century, remarked that "the occasional performance of circumcision" was probably "an imitation of Filipino custom."[24] In 1978, Jean T. Peterson reported that Agta fathers circumcised boys at puberty "without fanfare." The people explained that, although "no boy wants to be circumcised because he knows it will hurt," parents say that boys must submit "because they say so."

> When the feat is accomplished the boy is proud of his new manhood and glad that he obeyed his parents. Later, he may say that he wants to marry a certain girl; the parents may forbid it and insist he marry a girl of their choice. Recalling his dread of circumcision and eventual gratitude for it he will agree, knowing that his parents are perhaps aware of benefits which may result from an act which appears to offer only pain.[25]

Note the reference here to "new manhood" — indicating that, although relatively isolated geographically and culturally, the Agta are recognizably Filipino in their belief about what foreskin cutting accomplishes.

Foreskin Cutting in Muslim Communities

By the fifteenth century, Indonesian Muslim missionaries had reached the Sulu islands and probably Mindanao, while not long afterward Spanish Christian missionaries arrived in the central and northern regions. The result was that while most central and northern Filipinos are Christians, many in the south are Muslim. (However, Christian immigrants into Mindanao, seeking land and economic opportunity, have reduced Muslim dominance there.) Thus, whereas most Filipinos practice supercision, Muslims believe that circumcision is an essential requirement of Islam.[26] (One belief, however, unites almost all Filipinos: *some* kind of foreskin cutting is taken for granted as a step in the transition from boy to man.)

The religions of several societies in the Sulu Archipelago blend Islam with traditional beliefs and rituals. The Tausug, residents of Jolo and adjacent islands, adhere to the Shafi sect of Sunni Islam.[27] J. Franklin Ewing reported on their circumcision ceremony (*Mag-Islam*) in the 1950s. Boys aged about twelve are circumcised by the local Islamic religious leader (*imam*). The operation is usually held in the morning, accompanied by a celebratory feast.

> The boy squats over a mat, and two young men hold a white cloth over his head and wave it up and down during the ceremony. The Imam causes the foreskin to protrude through a split in a small piece of bamboo, and secures this position by slipping on a small rattan ring. He shears off the protruding part of the foreskin with a steel knife. After the operation the Imam applies powdered *dapaw niog* [not translated] to the wound. As he does so he utters the word *Bismallah* ("In the name of God") and binds it with a white cloth. Prayers are recited before and after the operation, the one subsequent to the circumcision being repeated after the Imam by the subject [i.e., the boy].
>
> At the end of the operation, the singing ceases. The abscised foreskin is placed in a half of a coconut shell and buried at the base of the house approach. Unless these practices are followed, the boy could anticipate trouble in getting married well, later on. After two days, the boy immerses himself in the sea and takes off the white cloth. If not well healed by that time, he goes bathing every day and applies fresh *dapaw niog* until he is healed. Then the boy is considered an adult Muslim.

Ewing also mentioned a "quasi-circumcision" of girls, performed by a midwife, lacking singing or feasting, witnessed only by women, and of "much less importance." He concluded that it involved rubbing a steel knife "over the anterior portion" of the labia, but without any cutting. Nevertheless, several people told him this was necessary to make the girl "a real Muslim." He believed that this practice — "obviously an imitation of male circumcision" — had "no parallel in the southern Philippines."[28]

However, another ethnographer, reporting on field work in the 1960s among the Yakan, a small Muslim group in Zamboanga (southwestern Mindanao) and the adjacent island of Basilan, stated that children of both genders are genitally cut — girls at about age three, boys between seven and twelve. The author commented that this (meaning both?) is "very common among the Indonesian Muslims."[29]

Children of both genders also undergo genital cutting among the Jama Mapun (or Bajau), a Muslim people living in several islands of the Sulu Archipelago and adjacent parts of Borneo. In 1967, Eric Casiño reported that boys are circumcised around age ten, in a rite attended only by male participants. The operation may be a unique version of "circumcision": "A sharp steel knife is used to cut a hole on the upper layer of the foreskin through which the glans is then inserted. The rest of the foreskin forms an irregular mass on the underside of the neck of the glans." Prayers are recited before and after the rite. Girls aged about seven have the clitoris "scratched several times with a bamboo knife." The operator is a male religious leader (*paki*), assisted by elderly women reciting prayers. Casiño commented that genital cutting of boys and girls is "regarded as an introduction of the growing individual into the Islamic community." The rites affirm "the social solidarity of the community by recalling their common institutional symbol of Islamization."[30]

In summary, despite the differences between Muslim societies and their northern neighbors, all believe that undergoing foreskin cutting is essential for a properly constituted man. Possibly supercision was the earlier southern practice, but circumcision replaced it when many accepted Islam. Notions about "cleanliness" are found here as elsewhere, but always the goal is sexual acceptability.

Foreskin Cutting and Real Men

A leading Filipino critic of foreskin cutting, Dr. Reynaldo Joson, has commented that the practice is "tradition-driven," supported only by custom.[31] But custom matters — certainly in this case, where custom is firmly connected with beliefs about male sexuality and desirability.[32] Many Filipino boys feel anxious and frightened about having their foreskins cut. But they know that everyone aspiring to social respectability must acquiesce; they know, too, that they need not undergo "baptism" (their term) alone but can shore up their courage by joining a group of boys their own age. The rewards are high: social acceptance and admission into the ranks of properly fledged young men, and the prospect of becoming a sexually desirable lover.

Beliefs about health and "hygiene" are indeed part of the picture. Underlying the general acceptance of cutting is the belief that an intact penis is dirty, odorous, and repellent. But in contrast to the United States, where one hears similar claims (though with most emphasis on supposed medical benefits), in the Philippines the foremost goal is a penis meeting standards of male fitness, respectability, and attractiveness. In short, Filipinos think of foreskin cutting not as *removal* for health benefits but as *improvement* to prepare for adult male identity and sexual desirability.

These attitudes, firmly engrained in individuals as they mature, are embedded in broader social considerations. An American Jesuit priest and anthropologist, long resident in the Philippines, described the high priority Filipinos place on "social acceptance." This is expressed in the Tagalog term *pakikisama*, which means getting along smoothly with everyone and "the lauded practice of yielding to the will of the leader or majority."

Children learn to be obedient and to accept social norms without resistance or complaint. Similarly, another author writing on "social customs" says that young Filipino children learn early "to internalize the important Filipino social value of *pakikisama* (to accompany or go along with for the benefit of group harmony), which serves as a guiding principle governing family relationships and interactions in the wider community." They learn that conflict should be avoided — that interpersonal relationships should be conducted smoothly.[33] A related term, *hiya*, describes the sense of shame experienced when one has violated social norms, or the fear of being shamed when even contemplating such a violation. *Hiya* has been defined psychologically as "a painful emotion arising from a relationship with an authority figure or with society, inhibiting self-assertion in a situation which is perceived as dangerous to one's ego." F. Landa Jocano adds that *hiya* is rooted in obligations inherent in the kinship relations that for many Filipinos constitute the context for most social interaction.[34] But in this case, social acceptability and sexual desirability are inseparable: *social* conformity yields *personal* rewards.

Intact boys and men, called *supot*, are said to have offensive odor, and it is generally understood that women disdain them as husbands or lovers.[35] A book on "understanding the Filipino," published in 1987 by a "Christian Literature Society" in the Philippines, portrays "circumcision" as the first phase of a "long-held tradition of initiation" called *binyag*, "baptism." (The second stage is first intercourse — either "pre-marital defloration" with a prostitute, or "where are no prostitutes," with a homosexual man — this in a book of "Christian literature"!) Circumcision, say these authors, is "still considered the flagship and test of manhood." Circumcised youths are "clean" and "much preferred by women." Intact boys are "not considered men yet"; they may be called "sissy" and may become "the object of a fair game for a lot of needling, taunting, and ribbing from both circumcised friends and enemies." Many village or town operations are carried out during "Holy Week," especially "Holy Saturday," the day before Easter, in the belief that "bleeding is not profuse during this [Lenten] season." The operation is performed "in a tree-covered backyard in the rural communities, or in an isolated shanty in the urban center."[36]

Two psychologists, writing in the 60s on "child rearing and personality development" in the Philippines, cited "circumcision" (as usual, probably meaning supercision) matter-of-factly, without commenting on its possible psychological significance:

> A boy in his preteens, some as early as nine if his friends are older than he is, starts to think of having himself circumcised. Sometimes the father makes arrangements, but usually there is a man in the village who does this as a favor for the boys in his neighborhood. The

> boy does not tell his younger siblings because they would not understand [but a nine- or ten-year-old would "understand"?] and would only tease him; however, his parents and older siblings usually know. His chores in the household are made lighter while he heals. The older members of the family recognize circumcision as his first step toward manhood. The boy begins to be called *binatilyo* (little man) instead of *bata* (child), and usually at seventeen, or even earlier depending on his evident maturity, he graduates into being *binata* (young man).[37]

The key term here seems to be *binatilyo*. The supercised boy enters a preliminary form of manhood; he becomes a "little man." The social implications seem quite distinct from those associated with our term "adolescent." Younger boys, still in the *bata* stage, are told nothing because that kind of knowledge would be inappropriate for someone not yet ready for "little man" status.

The study in Cebu City and Davao, mentioned earlier, described the experience of 114 males living in those two major urban areas. About half reported having been cut between ages ten and fourteen; most of the others had been younger, but six had been fifteen to eighteen. These were the top reasons given (in various combinations) for the operation: to avoid being called *supot* ("uncut"); part of tradition for a youth of the right age; to grow tall and physically fit; to keep penis free of smegma; to be able to impregnate a woman and have one's own children; told by parents to have this done; to be able to have a girlfriend and marry; because women prefer cut men for intercourse. Six respondents said it would ensure that their children would be "normal," while three thought it would enable them to "become intelligent."[38]

The researchers explained that the label *supot* has deep-seated negative connotations: To some of their respondents it meant that one has not only "a different looking *titi* [penis]" but "a cowardly character." Others said that the label "invites ridicule, shame, and embarrassment: someone is bound to shout at or call them by that name in public." Others linked intact status with being homosexual and lacking "courage to confront the pains, anxieties, and worries of circumcision." Most said that fear of being excluded from group activities and pressure to conform were much too strong to resist. They feared that women would "ridicule them" if they were intact.[39]

"College girls" interviewed around 1970 in Cebu City expressed strong preference for men who had been supercised. They referred to intact men as *pisut*, a term equivalent to *supot*, connoting "cowardliness" and "lack of cleanliness." One young woman said that if an uncut prospective fiancé resisted her "suggestion" that he be supercised, she would "decide to break up with him because I would find him a real coward, a *bayut*" ("sissy," translated here as "effeminate"). The women explained why they preferred supercised men: It's a "Christian custom" (but this seems to mean "civilized" or "refined" custom, not religious mandate); it makes a man "complete, a he-man, not a coward or sissy"; it's "good for the health"; it protects against "cancer of the genitals" (not mentioned in any other study I've seen); it "hastens physical growth"; it's necessary because uncut men are "dirty" and have a "bad odor"; it "helps in the sexual relationship of husband and wife" by providing "more pleasure and more satisfaction" — "because the operation makes 'it' bigger and fuller"; it's "the most natural thing" for a man, and "unnatural" for a man to be intact; it makes a man "stronger and more muscular." Regional physicians and traditional operators said that intact men from other provinces requested supercision to make them "more attractive to Cebu's discriminating women."[40]

These findings seem to be representative of what one might hear anywhere in the Philippines. Boys achieve and demonstrate manhood by agreeing to foreskin cutting, anticipating that otherwise women will reject them. Moreover, as I've noted, most Filipinos regard social conformity not as an unwelcome burden but as the essential foundation for pleasant, mutually satisfying social interaction.[41] It is not enough, however, to explain Filipino acceptance of foreskin cutting as acquiescence to authority and avoidance of social conflict. Filipino boys accept supercision or circumcision not simply to "conform" but because they truly believe that these procedures are a definitive gateway into manhood. Only men with cut foreskins are "real men." Speaking of these beliefs as entrenched custom or "social norm" is another way of emphasizing their depth and tenacity.

Mass Foreskin Cutting Events

Filipinos often conduct mass circumcisions known as "Operation Tuli" ("Operation Circumcision"). These are public, even festive, occasions in which dozens or even hundreds of boys are cut in a single day by visiting physicians (who perform as a public service!). Remarkably, some events are sponsored and underwritten by government organizations or personnel, or by Christian religious groups. As Jocano remarked, the mass cuttings are sometimes scheduled for "Holy Saturday" (also called "Sabado de Gloria" and "Black Saturday"), the day before Easter, because it is believed that the boys will bleed less that day.

One such "Operation Tuli" took place in June 2003 in a town on the island of Cebu. More than three hundred boys were supercised as a gratuitous public benefit. In that campaign nearly twelve hundred boys had already undergone the operation in nearby towns. Twelve physicians from the University of the Visayas, joined by a medical team from Armed Forces of the Philippines Central Command and by physicians from the local rural health unit, delivered the service — free of charge, thanks to the organizational talents of the local district representative and the generosity of the physicians. In addition, the representative distributed "pain relievers, vitamins, and antibiotic medicines" at a personal cost of P750,000 (about $15,000). His brother, the town mayor, boasted that the event "had given the boys substantial savings since a circumcision at a private hospital would cost them between P500 and P800" (about $10–$16). "Operation Tuli" in this town and neighboring communities is a "yearly project" of the representative.[42]

Ramos and Boyle, in their article on foreskin cutting in Batangas province (southwestern Luzon), say that hospitals "schedule mass circumcisions often including 100 or more boys." Physicians screen prospects in advance, to determine whether the foreskin has detached sufficiently to facilitate the surgery, and to examine for possibly disqualifying "congenital abnormalities." On the day of the surgery the boys are processed in assembly-line fashion.[43]

Boys sometimes organize mass foreskin cutting events on their own. I cited F. Landa Jocano's account of his own circumcision with four companions. But it appears that larger numbers may sometimes be involved. A March 2005 newspaper article described group events organized by boys in a town in Mindoro (the large island southwest of Luzon). Boys aged nine to twelve organized their own "rite of passage" to be conducted on "Black Saturday" ("because God is already alive [that day] so it would be less bloody"), with a traditional operator, not a physician. The operations were to be performed in a "forested area so girls cannot see them [because] if a girl sees the circumcision, the boys' scrotums will grow big." The operative technique appears to have been partial circumcision. Following cleansing with water and guava leaves, the operator made a hole in a white cloth, inserted the penis through the hole, and tied it with a piece of cloth — presumably to separate glans from protruding foreskin; then he sliced off the visible foreskin with a steel razor. Local people say that the penis has "a better appearance when circumcised the traditional way, compared to having it done in the hospital. ... The skin is stretched back, so it is still intact, unlike in a hospital, where the [remaining] skin is also cut off." A newly circumcised boy must not go out from Monday to Friday, "as it is risky and old folks claim that a snake will come out and bite him."[44]

Conclusion: Improving vs. Removing

Although some male Filipino infants are now being cut in hospitals, most boys submit when they are about eight to thirteen years old. No community, Christian or Muslim, specifies a precise age, but all agree that a boy must be cut before or soon after entering puberty. The universal justification is that being cut is an essential step along the path to becoming a complete adult — a "real man." The term "circumcision" may mean almost any form of foreskin cutting, usually supercision. Most operations involve removal of little or moderate amounts of foreskin tissue — certainly much less than is the case in American hospital circumcisions. Among Christians the usual operation is supercision — a dorsal cut with foreskin flaps folded down rather than removed.

Filipinos consider the primary purpose of foreskin cutting to be not *foreskin removal* but *penis modification*. Intact penises, they say, are ugly, dirty, and odorous; it is not the foreskin itself that is an object of scorn but the smegma beneath it. Filipinos seldom proffer medical or religious justifications for circumcision or supercision; instead they say simply that cut genitals are a requirement for adult social status and acceptance by women. Submitting to foreskin cutting is recognized as daunting and even frightening, but the promised rewards are irresistible.

Most Americans think of circumcision as a "hygienic" medical procedure that protects against an ever-changing assortment of illnesses. Some also think of the operation as cosmetic, but in a largely negative sense: avoidance of embarrassment in the presence of others. They think of foreskins as undesirable, even dangerous, and *remove* them. In contrast, Filipinos view both supercision and circumcision in a more positive light: they say it facilitates entry into manhood and improves sexual prospects; the purpose is *improvement.* Although extended consideration of this difference would lead us too far afield, it appears that attitudes to foreskin cutting correlate intimately with beliefs about manhood and sexuality. Campaigns against genital cutting in the Philippines (as everywhere) will benefit, therefore, from frank recognition of the power of culturally entrenched beliefs — particularly since they are inseparable from fundamental ideas about social maturation, gender identity, and sexuality.

References

1. I want to dedicate this introductory study to Dr. Yngve Hofvander, of Uppsala University, whose paper on "Violations Against Children in the Name of Religion and Tradition," delivered at the eighth international NOCIRC symposium in Padua, September 2004, stimulated me to think about foreskin cutting in nations other than my own.
2. Jocano FL. Growing Up in a Philippine Barrio. New York: Holt, Rinehart and Winston; 1969. pp. 57, 59 (p. 58 has an illustration of another subject). In a formal ethnography of this community, published the same year, Jocano says much less on this subject: "Among the men, the only ceremony [sic] which may possibly, though not necessarily, occur is circumcision." He notes that young boys are not cut "because a child does not wear trousers and his genitals are seen." Jocano doesn't say whether hospital physicians were circumcising frequently in the 1960s. Jocano FL. The Traditional World of Malitbog. Quezon City: Community Development Research Council, University of the Philippines; 1969. pp. 225–6. Although I have followed the author's lead here in calling the operation "circumcision," it was probably supercision. Foreskin cutting, most often supercision, is usually called circumcision by Filipino ethnographers and others.
3. Jocano FL. Growing Up in a Philippine Barrio. New York: Holt, Rinehart and Winston; 1969. p. 59.
4. Romeo B. Lee, Associate Professor of Behavioral Sciences at De La Salle University in Manila, estimates that 98% of Filipino males have been cut, most often by supercision. Personal communication, Dec. 6, 2005. I am indebted to Dr. Lee for kindly responding to my questions about the contemporary situation in the Philippines.
5. In an article published in 1936, a Filipino author reported seeing four methods of foreskin cutting, only one of which could be accurately labeled "circumcision." This method, called *tuli sa bao* (cutting with a coconut shell), involved use of a shell with a hole in the center, through which the prepuce was inserted, then sliced off with a razor. Post-operative treatment was extensive: A day later the wound was washed in a solution of boiled guava leaves. Then "fine scrapings of coconut shell" were applied, and the penis was held over "smoke produced by burning fresh jackfruit leaves." The wound was washed and dressed daily thereafter. Healing usually required "ten days to one month." Maceda GS. Some methods of circumcision in the Philippines. The Philippine Journal of Science 1936;58:513–7 [here, pp. 515–6]. Two other methods, variants of supercision, were *tuli sa gunting* (cutting with scissors), and *tuli sa itak* (cutting with a bolo, a Filipino machete), in which the prepuce was stretched over the blade and struck repeatedly with a banana leaf stalk until cut through. Finally, there was *tuli sa batakan*: stretching the foreskin over a piece of polished wood inserted into the ground and striking the dorsal surface with a knife until cut through. "In some instances," the author reported, "the operated person faints, and in this case the operator slaps his face to bring him back to consciousness, then applies the medicine and dresses the wound and the boy is considered

baptized" (pp. 514–5). Somewhat confusingly, Maceda also remarked that the method being employed by many Filipinos was "circumcision in the real sense performed by physicians," defined as "amputation" of the foreskin and "suturing of the free borders of the prepuce under local anesthesia" — presumably in hospitals or medical clinics (p. 513).

6. The essential reference is now Lee RB. Filipino experience of ritual male circumcision: knowledge and insights for anti-circumcision advocacy. Culture, Health and Sexuality 2006;8(3):225–34. In addition to this article, see: Ramos S, Boyle GJ. Ritual and medical circumcision among Filipino boys: evidence of post-traumatic stress disorder. In: Denniston GC, Hodges FM, Milos MF, editors. Understanding Circumcision: A Multi-Disciplinary Approach to a Multi-Dimensional Problem. New York: Kluwer Academic/Plenum Publishers; 2001. pp. 253–70; Alegre AD. Circumscribing Circumcision: Re-examining Routine Male Circumcision as a Violation of Human Rights. Thesis submitted to Ateneo de Manila University School of Law, 2004; and Lee RB, Norella LB. Between the Thighs: Penile Circumcision, Implants and Sexual Gadgets, 2002. [Available by title via Google.] The Ramos and Boyle article argues that foreskin cutting causes serious psychological damage to many Filipino boys and men. The Alegre publication is a dissertation by an attorney arguing against foreskin cutting on legal and ethical grounds. Lee and Norella are specialists in male sexuality research; their study and the 2006 article by Lee, cited above, are the most comprehensive available. I know of no publication specifically defending or advocating cutting — perhaps a reflection of the fact that there is little Filipino objection or challenge to the practice. For analysis of the sexual dimension of circumcision in British, and by extension American, culture, see Darby R. A Surgical Temptation: The Demonization of the Foreskin and the Rise of Circumcision in Britain.
Chicago: University of Chicago Press; 2005; and Glick LB. Marked in Your Flesh: Circumcision from Ancient Judea to Modern America. New York: Oxford University Press; 2005.
7. Alegre AD. Circumscribing Circumcision: Re-examining Male Circumcision as a Violation of Human Rights. Thesis submitted to Ateneo de Manila University School of Law, 2004, p. 65. I take the Tagalog spelling from Aspillera PS. Basic Tagalog. Rutland, VT: Charles E. Tuttle Co. 1969. pp. 215, 219.
8. An excellent survey is Jocano FL. Folk Christianity: A Preliminary Study of Conversion and Patterning of Christian Experience in the Philippines. Quezon City: Trinity Research Institute; 1981. He does not mention foreskin cutting.
9. Jocano FL. Filipino Prehistory: Rediscovering Precolonial Heritage. Manila: PUNLAD Research House; 1998.
10. A useful introduction is Steinberg DJ. The Philippines: A Singular and a Plural Place. 4th ed. Boulder, CO: Westview Press; 2000. pp. 53–77. See also his bibliography on historical research, pp. 237–44.

    To place this entire discussion in broader social and political context, it should be noted that many Filipinos, rural and urban, live in conditions of extreme poverty. The nation's economy is dominated by a small elite of business and professional people, while many of the rest sometimes face conditions harsh enough to mean near-starvation and death of infants and children. Filipinos are among the world's most numerous economic migrants, and many families are heavily dependent on remittances from emigrants to the United States and elsewhere. Anyone addressing the problem of foreskin cutting should recognize that social and economic problems may override all else in the minds of most Filipinos. Here is a brief summary of the situation by David Joel Steinberg (book cited above, p. 171): "The Philippines has one of the most unequal income distributions in the world. Some 42 percent of Filipino families in the urban areas and 58 percent of those in rural areas were living below the poverty line [in the 1980s]. Malnutrition increased throughout the 1980s. … From 1979 to 1989, an additional 12 million people were 'recruited into the ranks of the absolutely poor' [quoting a World Bank report]." See also Goodno JB. The Philippines: Land of Broken Promises. London and Atlantic Highlands, NJ: Zed Books; 1991 (Author's surname sic).
11. Scott WH. Barangay: Sixteenth-Century Philippine Culture and Society. Manila: Ateneo de Manila University Press; 1994. p. 25.
12. Scott's book includes a section titled "Mindanao and Luzon," but there is no mention of foreskin cutting. He does say (Barangay, p. 161) that "much that has been said of Visayan culture

in part 1 is also applicable to the coastal populations of Mindanao, and even to many communities in the interior."

13. Jane and James Ritchie provide an overview of supercision in Polynesia: "Traditionally Polynesian youths presented themselves with their age-mates for supercision. In 1938 the Beagleholes reported that on Tonga busloads of boys in gaily colored clothes trooped off to the local hospital, where the operation was performed with scissors under local anaesthetic, and then returned home for the feasting. Nowadays in Tonga it is done earlier but still with feasting. But both Holmes and Levy respectively for Samoa and Tahiti report that the operation is done by local practitioners with a razor blade, a bamboo knife, or piece of glass. In Western Samoa the operation is done shortly after birth. ... Almost everywhere the reason given for the operation is to promote cleanliness, but it is also a matter of some shame if it is not performed. In Tahiti, at least, some girls refuse to have intercourse with an unsupercised male. As a rule the requirement of supercision seems matter of fact rather than ceremonious, but where *lavalavas* [male skirts] may slip and some accidental exposure occur, one needs to be seen to be proper. Full circumcision, the complete exposure of the glans, is also improper — a man would only unsheath his penis to insult an observer." Ritchie J, Ritchie J. Growing Up in Polynesia. Sydney and London: George Allen and Unwin; 1979. p. 90. Two points worth special note: first, the obvious similarity to Philippine practice and attitudes; second, the feeling that complete exposure of the glans would be unseemly. The other ethnographies cited are as follows: Beaglehole E, Beaglehole P. Pangai: Village in Tonga. Wellington, NZ: Polynesian Society; 1941; Holmes LD. Samoan Village. New York: Holt, Rinehart and Winston; 1974; and Levy RI. Tahitians: Mind and Experience in the Society Islands. Chicago: University of Chicago Press; 1973. Levy's book on Tahiti includes extensive psychologically oriented discussion of supercision, pp. 117–22, 367–73, etc. Levy notes, p. 121, that eighteenth-century European voyagers to Tahiti stated that supercision was practiced then. See also Oliver D. Two Tahitian Villages: A Study in Comparison. Provo, UT: Brigham Young University Press; 1981. pp. 367–8. Other accounts include a detailed discussion of supercision in Tikopia, a Polynesian outlier, in Firth R. We, the Tikopia. 2nd ed. Boston: Beacon Press; 1963, esp. pp. 382–92 and 427–33; and Buck PH (Te Rangi Hiroa). Mangaian Society. Honolulu: Bernice P. Bishop Museum Bulletin 122; 1934. pp. 89–90. For a description of supercision in Sulawesi (formerly Celebes), see Hollan DW, Wellenkamp JC. The Thread of Life: Toraja Reflections on the Life Cycle. Honolulu: University of Hawaii Press; 1996. pp. 62–9. See also Hill TH, Budiharsana M. Male circumcision and penis enhancement in Southeast Asia: matters of pain and pleasure. Reproductive Health Matters 2001;9(18):60–7. Authoritative studies of Pacific prehistory by Peter Bellwood do not mention foreskin cutting. Bellwood P. Man's Conquest of the Pacific: The Prehistory of Southeast Asia and Oceania. New York: Oxford University Press; 1979; and Prehistory of the Indo-Malaysian Archipelago. Sydney: Academic Press; 1985.
14. Lee RB, Norella LB. Between the Thighs: Penile Circumcision, Implants, and Sexual Gadgets; 2002. pp. 20–4; drawings by respondents on p. 22 and by authors on p. 24. (Some drawings by respondents seem to portray partial but not inconsiderable foreskin removal.) Available via Google as "Between the Thighs." (Note that page numbers in the printed text differ from those online; my references are to printed text.) In a recent study conducted in Batangas province (southwestern Luzon) among 3253 boys aged 11 to 16, Ramos and Boyle described three kinds of incision: a dorsal slit (supercision); a "V-cut" with "two intersecting diagonal cuts" and removal of the intervening tissue; and a "German cut" (because the removed tissue is said to resemble a World War II German helmet), in which the foremost tissue is removed by a horizontal incision. Ramos S, Boyle GJ. Ritual and medical circumcision among Filipino boys: evidence of post-traumatic stress disorder. In: Denniston GC, Hodges FM, Milos MF, editors. Understanding Circumcision: A Multi-Disciplinary Approach to a Multi-Dimensional Problem. New York: Kluwer Academic/Plenum Publishers; 2001. p. 256. This is the only mention of "V-cut" I've encountered.
15. Lee RB, Norella LB. Between the Thighs: Penile Circumcision, Implants and Sexual Gadgets; 2002: p. 26. Available as "Between the Thighs" via Google. The final line reads "the circumcised then dressed ...," but I think the first d is a typo.

16. Rubel AJ, Liu WT, Brandewie E. Genital mutilation and adult role behavior among lowland and Christian Filipinos of Cebu. American Anthropologist 1971;73:806–10 [here, p. 807].
17. Jocano identifies the Panay villagers as Hiligaynons (an ethnolinguistic term). Here is what happens in a Hiligaynon supercision: The instruments are "a stainless steel knife and a guava branch carved into two different shapes, one L-shaped and the other rectangular." With the "patient" seated, the operator places "the tip of the penis beneath the end of the L-shaped wood," pushes the foreskin forward until the wood is visible, and makes certain that "the skin is stretched on the wood to avoid cutting the veins." Then he "places the sharper blade of the knife on top of the skin" and strikes it "with the rectangular wood until the skin is cut ... leaving the tip of the penis bare." Note that he does not actually remove tissue. After application of merthiolate, penicillin ointment, or chewed guava leaves, the wound is bandaged, and the circumciser himself changes and washes this daily. The newly cut youth must refrain from eating tomatoes, pork, and dried fish. Also, coming near a menstruating woman may cause "bleeding and swelling" of his wound. Jocano FL. The Hiligaynons: An Ethnography of Family and Community Life in Western Bisayas Region. Quezon City: University of the Philippines Press; 1983. pp. 193—4. In the personal narrative quoted in my introduction, Jocano does not identify his own town as ethnically Hiligaynon, but it probably is.
18. Manuel EA. Manuvu' Social Organization. Quezon City: University of the Philippines Press; 1973. p. 96.
19. Rebollido RG. Passage to Manhood. General Santos, P.I.: Sun Star; March 21, 2005. http://www.cirp.org/news/sunstar03-21-05/ (June 23, 2005).
20. Jocano FL. The Ilocanos: An Ethnography of Family and Community Life in the Ilocos Region. Quezon City: Asian Center, University of the Philippines. 1982. pp. 159–60. Nydegger WF, Nydegger CN. Tarong: An Ilocos Barrio in the Philippines. In: Whiting B, editor, Six Cultures. New York: John Wiley & Son; 1966, is a brief ethnography, based on research by two anthropologists in 1954, in an Ilocano community in coastal northwestern Luzon. Although specifically designed to obtain information on psychological dimensions of child-rearing practices, this study does not mention foreskin cutting. Nor is there any mention in Griffiths S. Emigrants, Entrepreneurs, and Evil Spirits. Honolulu: University of Hawaii Press; 1988, another brief account of an Ilocano community conducted in 1973. Since supercisions may be performed intermittently, it's possible that none occurred while these ethnographers were in Ilocano communities; or perhaps they didn't find the subject worth mentioning.
21. LeBar FM, editor, Ethnic Groups of Insular Southeast Asia. Volume 2: Philippines and Formosa. New Haven, CT: Human Relations Area Files Press; 1975. p. 99. This statement is based on a 1954 article by Morice Vanoverbergh (cited on p. 161), which I have not seen.
22. Ethnographers vary, even in their own publications, with regard to use of the singular or plural form — e.g., the Kalinga, the Kalingas. On the Ilongot, see Rosaldo MZ. Knowledge and Passion: Ilongot Notions of Self and Social Life. Cambridge, Eng.: Cambridge University Press; 1980, and Rosaldo R. Ilongot Headhunting 1883–1974: A Study in Society and History. Stanford, CA; Stanford University Press; 1980. The Ilongot are not to be confused with the lowland Ilocano, discussed earlier. A detailed ethnographic study of the Tinguian, conducted by an experienced anthropologist in 1907–1908, does not mention genital cutting: Cole F-C. The Tinguian: Social, Religious, and Economic Life of a Philippine Tribe. Chicago: Field Museum of Natural History, Publication 209; Anthropological Series 14(2);1922. A record of the autobiographies of three Ifugao subjects, including two men in their forties or fifties, does not mention genital cutting: Barton RF. Autobiographies of Three Pagans in the Philippines. New Hyde Park, NY: University Books; 1963 (originally published in England in 1938 as Philippine Pagans).
23. Dozier EP. Mountain Arbiters: The Changing Life of a Philippine Hill People. Tucson: University of Arizona Press; 1966. p. 97; see also Dozier EP. The Kalinga of Northern Luzon, Philippines. New York: Holt, Rinehart, and Winston; 1967. pp. 40–1. Barton RF. The Kalingas. Chicago: University of Chicago Press; 1949, based on fieldwork in 1916 and 1941, does not mention genital cutting.

24. Garvan JM. The Negritos of the Philippines. H. Hochegger, editor. Vienna: Ferdinand Berger Verlag; 1964. pp. 82, 87.
25. Peterson JT. The Ecology of Social Boundaries: Agta Foragers of the Philippines. Urbana: University of Illinois Press; 1978. pp. 56–7.
26. It is well known that there is no mention of circumcision in the Koran, but most Muslims throughout the world consider the practice mandatory for pre-pubertal boys.
27. Basic information on the Tausug and all other ethnically distinctive societies discussed in the next two sections is available in LeBar FM, editor, Ethnic Groups of Insular Southeast Asia. Volume II: Philippines and Formosa. New Haven, CT: Human Relations Area Files; 1975.
28. Ewing JF. Some rites of passage among the Tawsug [sic] of the Philippines. Anthropological Quarterly 1958;31(2):33–41 [here, p. 36–8]. (Reprinted with minor editorial changes in Gowing PG, McAmis RD, editors, The Muslim Filipinos. Manila: Solidaridad Publishing; 1974. pp. 132–7.)
29. Wulff I. Features of Yakan culture. In: Gowing PG, McAmis RD, editors, The Muslim Filipinos. Manila: Solidaridad Publishing; 1974. pp. 242–58 [here, p. 252]. (Originally published in a Danish journal: Folk 1965;6.)
30. Casiño ES. Folk Islam in the life cycle of the Jama Mapun. In: Gowing PG, McAmis RD, editors, The Muslim Filipinos. Manila: Solidaridad Publishing; 1974. pp. 165–80 [here, p. 170]. (Originally published in Philippine Sociological Review 1967;15:34–48.) Edward Kasman, writing in the 1960s, described genital cutting (of boys only) among the Samals, a linguistic term covering various ethnic groups in the Sulu Archipelago. While older Samals were farmers, fishermen, and boat-builders, members of the younger generation had become acculturated to modern life, practicing as teachers, professionals, and merchants. Nevertheless, Samal culture includes a variety of folk medical beliefs and practices. Kasman described circumcision as a "simple ritual" (here called *Pag-Islam*), performed on children aged seven to ten, so that they may be taken "into the fold of the Islam faith." Without circumcision it would be difficult for a young man to marry.

    On the day appointed for the circumcision, the boy is asked to take a bath. There is no lavish preparation and the simple rite takes place between just two persons, the *pangingislam* ("circumciser") and the boy. … [The boy] strips himself and has only a bedsheet about his loins. He sits on the floor and extends and parts his legs to give working room to the *imam*.

    The priest sits directly in front of the boy and takes the foreskin of the penis with two splits of bamboo, and with a small knife he cuts a sizeable portion of the foreskin off. The cutting of foreskin is accompanied with prayer. Then the priest places a coconut shell under the penis to catch the blood dropping from the wound. The *imam* treats the wound with powdered coffee and wraps the wounded part of the organ with a white piece of cloth. The organ generally swells after the operation.

    The operator receives rice, coconuts, and a small monetary payment. The boy is responsible thereafter for his own care. He must avoid chicken droppings, and must not walk over a rice pestle to discourage swelling. Later he bathes in the sea and removes the cloth; then he applies fresh coffee powder and rewraps his penis with new cloth. When he heals he is declared to be "a real Muslim." Kasman ES. Folk medicine and health practices among the Sulu Samals. Solidarity 1969;4:44–51 [here, p. 48].
31. "Question and Answer: Dr. Reynaldo Joson on Project Xtulepinoy 2003." http://xtulepinoy.tripod.com/qacircumcisionrjo3.htm (July 6, 2005). Available via Google at "Reynaldo Joson."
32. Lee RB. Filipino experience of ritual male circumcision: knowledge and insights for anti-circumcision advocacy. Culture, Health and Sexuality 2006;8(3):225–34. On foreskin cutting as social norm, see Waldeck SE. Using male circumcision to understand social norms as multipliers. University of Cincinnati Law Review 2003;72(2):455–526.
33. Lynch F. Social acceptance reconsidered. In: Yengoyan AA, Makil PQ, editors, Philippine Society and the Individual: Selected Essays of Frank Lynch, 1949–1976. Michigan Papers on South and Southeast Asia 24. Ann Arbor, MI: Center for South and Southeast Asian Studies, University of Michigan; 1984. pp. 23–91. Rodell P. Culture and Customs of the Philippines. Westport, CT: Greenwood Press; 2002. pp. 196–7. Lynch concluded that conformity seemed especially important for the population comprising "rural, lower class, poorly educated, tradi-

tional, employees, men": p. 77. Rodell does not make this distinction; nor does DJ Steinberg, who simply says that *pakikisama* is "one of the salient characteristics of Philippine life": Steinberg DJ. The Philippines: A Singular and a Plural Place. 4th ed. Boulder, CO: Westview Press; 2000. p. 4.

34. Jocano FL. Filipino social structure and value system. In: Jocano FL, editor, Filipino Social Structure and Value Orientation: Filipino Cultural Heritage. Manila: Philippine Women's University Graduate School; 1966. pp. 1–26; and Bulatao J. The 'hiya' system in Filipino culture. In: Jocano FL, editor, Filipino Social Structure and Value Orientation: Filipino Cultural Heritage; 1966. pp. 27–40 [here, p. 28]. See also comments on how sensitivity to *hiya* predisposes boys and adults to accept foreskin cutting as natural and normal: Ramos S, Boyle GB. Ritual and medical circumcision among Filipino boys: evidence of post-traumatic stress disorder. In: Denniston GC, Hodges FM, Milos MF, editors. Understanding Circumcision: A Multi-Disciplinary Approach to a Multi-Dimensional Problem. New York: Kluwer Academic/Plenum Publishers; 2001. pp. 253–70 [here, p. 254]. Jocano presents a detailed analysis of these terms and others in Jocano FL. Filipino Value System: A Cultural Definition. Manila: PUNLAD Research House; 1997.
35. Ramos and Boyle note that *supot*, defined in dictionaries simply as genitally intact, "originally meant constricted or tight," and in colloquial usage can mean "homosexual." Ramos S, Boyle GB. Ritual and medical circumcision among Filipino boys: evidence of post-traumatic stress disorder. In: Denniston GC, Hodges FM, Milos MF, editors. Understanding Circumcision: A Multi-Disciplinary Approach to a Multi-Dimensional Problem. New York: Kluwer Academic/Plenum Publishers; 2001: 253–70 [here, p. 255].
36. Andres TD, Ilada-Andres PCB. Understanding the Filipino. Quezon City: New Day Publishers; 1987: pp. 85—6.
37. Guthrie GM, Jacobs PJ. Child Rearing and Personality Development in the Philippines. University Park, PA: Pennsylvania State University Press; 1966: p. 140. The casual attitude expressed here certainly contrasts with the conclusions of Ramos and Boyle.
38. Lee RB, Norella LB. Between the Thighs: Penile Circumcision, Implants and Sexual Gadgets; 2002. pp. 18–20; available online via Google as "Between the Thighs." (Note that pagination in the report, at bottom right of page and cited here, will probably differ from pagination online.) This report includes information on how "providers" (physicians and traditional operators) decided whether or not to cut. Of the twelve interviewed, six or fewer offered each of the following answers: the penis should be palos, with a readily retractable foreskin; the "client" should be ready "to undergo the experience" and of proper age (not defined); the penis should be free of "abnormality"; parents should consent. Traditional operators were especially attentive to location of veins.
39. Lee RB, Norella LB. Between the Thighs: Penile Circumcision, Implants and Sexual Gadgets; 2002: p. 18; available online via Google as "Between the Thighs." (See note 38 regarding pagination.)
40. Rubel AJ, Liu WT, Brandewie E. Genital mutilation and adult role behavior among lowland and Christian Filipinos of Cebu. American Anthropologist 1971;73(3):806–10 [here, p. 807]. The answers referring to health and cancer suggest that educated Filipinos may be more aware of American-style medical beliefs. These researchers were testing a hypothesis advanced by Harrington that circumcision was associated with societies having prominent sex role differentiation, while supercision was associated with minimal differentiation. Their sample, added to Harrington's and another unpublished report on supercision in central Luzon, yielded eight societies, of which six conformed to the hypothesis. Harrington C. Sexual differentiation in socialization and some male genital mutilations. American Anthropologist 1968;70(5):951–6.
41. Writing on Filipino "culture and customs," Paul Rodell says that children learn early that cooperativeness, conformity, and deference to elders are essential for social harmony. Hence, "Filipino youth soon take on the approved values of the wider community … [particularly] "a sense of obligation … and a dynamic, ongoing practice of reciprocity. … An individual who will not or cannot get along with the group suffers the worst fate possible: being cast out of the group." Rodell P. Culture and Customs of the Philippines. Westport, CT: Greenwood Press; 2002: pp. 196–8.

42. Gullas' Operation Tuli Holds Last Leg in Talisay. Cebu City: The Freeman, June 1, 2003. Available at www.cirp.org/news/thefreeman06-01-03/ (June 23, 2005). The representative is named Jose Gullas.
43. Ramos S, Boyle GJ. Ritual and medical circumcision among Filipino boys: evidence of post-traumatic stress disorder. In: Denniston GC, Hodges FM, Milos MF, editors. Understanding Circumcision: A Multi-Disciplinary Approach to a Multi-Dimensional Problem. New York: Kluwer Academic/Plenum Publishers; 2001: 253–70 [here, p. 255]. In an ethnography of a peasant community in Panay, published in 1976, F. Landa Jocano included two sentences on circumcision: "Most males are circumcised when they are well advanced in adolescence. Other informants report that they were circumcised in the hospital shortly after they were born." Jocano FL. Tuburan: A Case Study of Adaptation and Peasant Life in a Bisayan Barrio. Quezon City: Capitol Publishing, 1976: p. 110. Two points to note: first, this was a small "peasant" community, not an urban center; second, although hospital circumcision (or supercision?) was still not the most common procedure, it was happening often enough to be mentioned.

    Romeo B. Lee says that upper and middle class boys "tend to have their circumcision performed in hospitals and at times during infancy." He reports that physicians "perform circumcision in clinics and hospitals, and for small clinics in particular, you would find advertisements for circumcision services posted at the front door." Personal communication, Dec., 6, 2005.
44. Virola MT. Boys' Rite of Passage Best on Black Saturday. Makati City: Philippine Daily Inquirer, March 20, 2005. Available at www.cirp.org/news/daily inquirer03-20-05/ (June 20, 2005).

    Opposition to foreskin cutting is developing in the Philippines, albeit slowly. In the medical community, the leader has been Dr. Reynaldo Joson, a prominent Manila surgeon, who says that in 1999 he realized the "senselessness of this practice" and its attendant dangers, and began campaigning against it. His personal recollection illustrates the power of the very tradition he opposes, since he continued to perform "circumcisions" for two more years, and even now accepts its appropriateness for "therapeutic" reasons.

    In 2001, I completely stopped doing non-therapeutic circumcision, after I advised and convinced my son not to go for the tradition-driven circumcision. I felt I should not have double standards, i.e., not doing circumcision on my own son and yet doing circumcision on patients other than my son. I remember vividly I turned down a request from a surgeon-colleague to do circumcision for his son that year.

    As part of his personal campaign, Joson successfully persuaded his immediate colleagues and a number in other Philippine hospitals to "make a stand on No to routine circumcision and No to Operation Tule" (sic) — the latter defined as "performance of tradition-driven, non-therapeutic circumcision in adolescents." Asked about possible religious basis for cutting, he replied that since most Filipinos are Catholics and the Church "not requiring circumcision for salvation," religion "is not a strong contributory factor for the rampant practice of circumcision in the Philippines."

    How does this pioneer intend to reduce or end "tradition-driven non-therapeutic circumcision"? Joson proposed that through "public health education and advocacy" he might change parental beliefs, "abolish the peer pressure," and convince physicians that routine circumcision should end. He seemed to anticipate most resistance from physicians "who have already acquired the habit of doing the procedure left and right and those who tend to ride on all traditions to avoid any conflict." Education would require publicizing a basic message: "*Tule: Hindi Na Kailangan! Masakit Pa!*" (Circumcision Is Not Necessary! It's Painful!). Joson RJ. Question and Answer: Dr. Reynaldo Joson on Project Xtulepinoy 2003. http://xtulepinoy.tripod.com/qacircumcisionrj03.htm (July 6, 2005). Dr Joson said that he approves of "therapeutic circumcision" for phimosis (defined as "restrictive foreskin"), paraphimosis, and balanitis.

    Commenting on contemporary efforts to reduce foreskin cutting, Romeo B. Lee says that "the campaign's arguments which are based on the US' medical primacy are not culturally appropriate." Personal communication, Dec. 6, 2005.

Filipino Male Experience of Ritual Circumcision

Perspectives and Insights for Philippine-Based Anti-Circumcision Advocacy

Romeo B. Lee

Abstract Male circumcision is a well-publicized phenomenon, but much of what is known at the international level concerns the West's neonatal medical circumcision. This report offers information instead on the Philippines' ritual prepubescent circumcision. While the report addresses the aforementioned information gap, its perspectives and insights are likewise intended to inform the Philippine-based anti-circumcision campaign of the need to ground its arguments on the culture within which ritual circumcision exists. Currently, the advocacy is anchored on arguments derived from the Western male experience of medical circumcision. Data in this report were derived from a 2002 Philippine circumcision study—a component in a Southeast Asian research on genital enhancement practices—whose aim was to form evidence-based perspectives with advocacy purpose. The study interviewed 114 circumcised Filipino males, of varying ages, who were selected purposively. The report highlights the traditional character of Filipino men's ritual circumcision experience. Among others, it underscores the links of the ritual phenomenon with reasons of masculinity, which are hastened by broader community involvement. The report offers broad strategies to making the anti-circumcision campaign culturally appropriate.

The Philippines has been one of the world's major circumcising societies. Almost all Filipino males — the estimate is that more than 90% of them[1] — are circumcised. It is unknown how a phenomenon with prehistoric roots[2] has been transformed into the country's most enduring and universal tradition for males. But what is known, albeit in a highly restricted way, is that circumcision is currently being advocated in the country as an unnecessary practice. The arguments for its non-necessity are anchored on three grounds: its medical effectiveness (for instance, against urinary tract and HIV infections, penile cancer, balanoposthitis, and phimosis) is by no means certain; it is a painful and risky procedure with attendant complications; and, when undertaken without informed consent, it comprises an infraction of bioethics and human rights measures (http://www.xtulepinoy.tripod.com).

Behavioral Sciences, De La Salle University-Manila, Manila, Philippines

G. Denniston et al. (eds.), *Circumcision and Human Rights*,

I argue that the campaign's three-pronged arguments, albeit sound according to the experience of North American males of medical circumcision, are not apt for use in the ongoing Philippine-based anti-circumcision advocacy because the Filipino male experience pertains to a phenomenon whose existence and persistence have rested not on medical but on a ritualistic impetus. It is a way through which a Filipino secures and achieves his masculinity.[3–5] Although known in a broad sense, the linkage of the Philippines' circumcision with the manhood issue is only scarcely understood at present. For instance, details on how Filipino men, particularly those circumcised, interpret the reasons for their submission to the procedure are lacking. What reasons do these men hold and how do they mirror, in specific terms, the masculinity impetus? Answers to this question are pertinent: if the Philippine advocacy were to be re-directed towards becoming culturally grounded, it requires detailed information about the essence upon which the country's ritualistic circumcision is embedded. In fact, for the advocacy to be holistically and substantively anchored on the culture of Philippine circumcision, its information needs should focus not merely on why the tradition is being observed but on how it is experienced by Filipino men as a whole, and, more importantly, on how it is performed. Provision is as critical an aspect as adoption — information about it would shed additional light on the tradition's persistence, which also would have implications for the anti-circumcision campaign's cultural grounding.

This paper discusses detailed empirical information on the adoption and provision of the Philippines' ritualistic circumcision. Data were derived from a circumcision study (henceforth referred to as CIRCS) undertaken in 2002, involving personal interviews with 114 circumcised males of varying ages and 12 circumcisers (6 medical professionals (doctors, midwives, and nurses) and 6 traditional, non-medical circumcisers).

Respondents were recruited using snowball and referral techniques. Respondents were recruited from three urban communities and three of their neighboring semi-rural or rural communities; the three sets are geographically distant from each other. Before their interviews, all respondents were informed of the study's objectives and importance, and of the confidentiality of their reported information. Consent, of the minors' parents and adult respondents, was sought prior to interview. To date, the Philippines lacks a formal ethical review system.

The 114 circumcised males were asked about the timing of and reasons for their circumcision, their providers and the type and cost of their procedure, complications (during and after circumcision), and the general effects of circumcision. The 12 circumcisers were queried about their clients and the cost, design, procedure, and complications related to their conduct of circumcision. Interviews were conducted in the Philippine languages of Cebuano or Tagalog (the report's data represent their English translations). Interviews lasted for about an hour and were conducted in offices, community centers, and vacant lots with audio privacy. Respondents were offered a modest incentive (a grocery bag with about US$3 worth of goods). The data presented in this paper were analyzed using frequency count.

Results

Adoption of Circumcision

The 114 circumcised males interviewed were aged between 13 and 51 (mean 25.9), two-thirds of whom were single, while others were married, living in, or separated. The majority (59.7%) was employed (laborers, drivers, construction workers, repairmen, security guards, welders, plumbers, painters, electricians, and vendors), with the remainder being unemployed and/or students. All were Catholic.

Half of the respondents (51.7%) were circumcised at ages 10–14, while others when they were 5–9 or 15–18 years old. When asked, respondents had several explanations as to why they had circumcision (Table 1). Five of these stood out as the most frequently mentioned reasons: they wanted to avoid being teased by peers as being *supot* (uncircumcised) (66.7%); they perceived it as part of growing up, a tradition for boys who have reached the right age (41.2%); they desired to grow tall and physically fit (29.8%); they wanted to free their penis of dirty and foul smelling *kupal* (smegma) (22.8%); and they desired to cause pregnancy and to sire a child (22.8%). The other, less commonly mentioned reasons included respondents' belief that they needed the procedure so that they could have a girlfriend and get married; to enhance the form of their penis and size of their penile glans; and to facilitate the entry of their penis during vaginal intercourse. Some stated that they were circumcised because women preferred a circumcised partner; a handful cited religion as their motivation.

Two of every three respondents (68.4% of 114) were circumcised by a lay or traditional provider, recalled as their friend, a relative, or a village official. Other respondents (31.6%) were circumcised by a medical doctor, midwife, or nurse.

Table 1 Respondents' reported reasons for circumcision[a]

Responses	No	%
1. To avoid being teased by peers as *supot* (uncircumcised)	76	66.7
2. Already a grown up, of the right age — part of the tradition for a boy like him to undergo circumcision	47	41.2
3. To grow tall and physically fit	34	29.8
4. Wanted his penis to be free of smegma	26	22.8
5. To be able to cause pregnancy, and wanted to have own child	23	20.2
6. Parents told him to undergo the procedure	21	18.4
7. It was a requirement to court a girl, to have a girlfriend, and to get married	14	12.3
8. Women like to have sexual intercourse with a circumcised Partner	12	10.5
9. To facilitate entry of his penis during sexual intercourse	7	6.1
10. To enhance the form of his penis and to make his glans larger	7	6.1
11. It is in the Bible that a Christian must be circumcised	4	3.5
12. To become intelligent	3	2.6
13. Circumcision was free	2	1.8

[a] Multiple response (n = 114).

Regardless of provider type, all respondents had a dorsal slit "circumcision": the prepuce was cut lengthwise (without excision) and then folded on the sides (with suturing and anesthesia, if performed medically, or without suturing, if done traditionally). While the majority (57.9%) paid for their circumcision (in cash, US$ 0.90 for traditional providers and 100–200 pesos for medical professionals or in kind, a pack of cigarettes, a bottle of beer, or snacks), others (36%) had free circumcision, courtesy of socio-civic and medical organizations and politicians.

Two-thirds of respondents (64.5%) assessed their circumcision as physically painful. Among those who received a medical procedure, they experienced pain when they were injected with the anesthesia and during the suturing of their cut prepuce. Among those traditionally circumcised, pain was experienced during the insertion of the piece of wood into their prepuce and the cutting of the prepuce without anesthesia. No single respondent mentioned that they had complications at the time of their circumcision. However, insofar as post-circumcision complications were concerned, the majority (59.6% or 68) — regardless of whether the procedure was medical or traditional — had a swollen and inflamed penis. Four respondents, in addition, reported having profuse bleeding and pus in their sex organ. In addressing their complications, two had sought advice from their circumciser, while six did not do anything. Almost all of those with complications (60 of 68) self-medicated — for example, by dousing their penis with juice from boiled guava leaves; bathing in the sea or river in the early hours of the day regularly for 3 days after circumcision; applying alcohol, Betadine, or penicillin to the circumcised penis; taking antibiotics and pain killer tablets; by applying sawdust or reddish brown grains found at the base of the leaf stem of the coconut tree.

In terms of the overall effects of circumcision, one in every 10 respondents (11.4%) said that the procedure had not had any effect on them. However, most (88.6% or 101 of 114) said the opposite, stating that their circumcision had several effects on them (Table 2). Three were most mentioned: that their friends had

Table 2 Perceived effects of circumcision[a]

Responses	No	%
1. None	13	11.4
2. Friends had stopped teasing him *supot* (uncircumcised)	55	54.5
3. It feels good knowing that one's penis is free of smegma	30	29.7
4. Partners are pleased because the glans is exposed, hence sexual penetration is easy	30	29.7
5. Now that his glans is exposed, he likes the sensation during sexual penetration, and thus easier for him to have orgasm	14	13.8
6. He has grown tall, his voice changed, and his body is healthy	11	10.9
7. He is now comfortable and confident knowing that his sexual partner would not laugh at him for being uncircumcised	11	10.9
8. He has caused pregnancy and he now has children	5	4.9
9. He frequently masturbates and desires intercourse	5	4.9
10. It is now easy to find sexual partners	3	2.9
11. He feels good that he followed a Christian tradition	1	0.9

[a] Multiple response (n = 114).

stopped teasing them as uncircumcised (54.5%); that they had felt good realizing that their penis was forever free of smegma (29.7%); or that they had pleased their partners given that, with their exposed glans, sexual penetration had been easy (29.7%). Respondents also reported that as a result of the procedure, they had felt and come to like the penile sensation during sexual penetration, thus orgasm is now easier for them (13.8%); they had become tall, had a changed voice, and had a healthy body (10.9%); they had felt more comfortable and confident knowing that their sexual partners would never laugh at them anymore for being uncircumcised (10.9%). The religious dimension of circumcision was rarely mentioned.

Provision of Circumcision

The dozen circumcisers, six medical professionals (MPs) and six traditional providers (TPs), had been conducting the practice for more than two decades (mean: 21.3 and median: 21.5). MPs became circumcisers because the procedure was an integral part of their medical services when they established their own clinic, when they joined a clinic, or because people had asked them to perform it. On the other hand, the TPs took the role because it was handed down to them by their relatives (grandfathers and fathers) and neighbor-friends, or upon the request of some teenagers who wanted to undergo the tradition.

For all circumcisers, Filipino boys submit to the procedure because it is a rite of passage; a way to free one's penis from smegma; to avoid being teased by friends as uncircumcised; or due to the perception that women prefer a circumcised partner or that it is a pre-requisite for school enrollment (Table 3).

Respondents reported that the number of their clients varies accordingly in a year. However, they said that the number of teenage boys wanting circumcision surges during summer (April–May), the period of school vacation. During this season, respondents would perform the procedure on between 15 and 40 boys a month. Outside of this period, respondents estimated they would have 2–10 clients a week at most, and 1–3 clients a month at least.

Respondents reported that the number of their clients in a year varies according to season. However, they said that the number of teenage boys wanting circumcision increases during summer (April–May), which is the period of school vacation.

Table 3 Circumcisers' stated reasons why boys undergo circumcision[a]

Responses	No
1. It is a rite of passage — part of growing up and a social and cultural norm and expectation	12
2. For hygiene purposes — so that their penis is free of smegma	8
3. Ridiculed or teased by friends of being *supot* (uncircumcised)	6
4. Women prefer circumcised men	1
5. It is a requirement for school enrollment	1

[a] Multiple response (n = 12).

Table 4 Factors that circumcisers considered before circumcising boys[a]

Responses	No
1. The penis should be *palos* already (retractable)	7
2. Structural abnormality in one's penis (e.g., too many veins)	7
3. Must be ready for it — relaxed and not agitated	6
4. Age (at least 10)	6
5. Parental consent and presence	4

[a] Multiple response (n = 10).

During this season, respondents would circumcise 15–40 boys a month; beyond this period, the number would be considerably lower. Their clients are aged between 5 and 14 years. MPs would charge their clients US$4–30, in contrast to TPs' US$0.20–1.00 per client. Circumcisers mostly perform the dorsal slit.

Before circumcising their clients, respondents — except for two TPs — would take into account one or more factors (Table 4). They made sure that the boy's penis is already *palos* (retractable) and, also, that it bears no abnormality; that the boy himself should be ready for the procedure, and likewise of the "right" age; and lastly, that there is parental consent and presence.

To the circumcisers, inflammation is the most typical problem faced by the newly circumcised individuals, with penile bleeding and presence of pus in the organ considered as less common complications. The dozen circumcisers interviewed had varied reports as to the complications experienced by their own clients. While some respondents had not had any client with a problem, others had a few clients with a problem or two, and more than half had clients reporting several complications. Respondents cautioned that their reports on the matter are inaccurate, given that they lack a systematic means of ascertaining how many and what problems their clients had encountered.

Discussion

Although male circumcision is a common tradition in several countries throughout the world, the prevailing reasons and justifications for its continuing practice are not the same; they, in fact, vary from one culture to another. For instance, the North American circumcision is a medical phenomenon and, therefore, anchored on medical imperatives, while as a ritualistic tradition, the Philippine circumcision is pursued for masculinity reasons. The Philippine-based anti-circumcision advocacy should be cognizant of the stark and marked differences between these two circumcising cultures and, thus, should refrain from using the North American's medical arguments for its local work. Instead, it first needs to examine the unique culture within which Philippine circumcision exists and persists, from which it can then form insights and perspectives through which to craft and implement a culturally grounded campaign. The CIRCS data discussed earlier represent an initial set of information that can help inform the grounding.

There is little doubt that the cultural grounding of the local campaign should focus on Filipino masculinity for its theme.[6] The CIRCS evidence detailed that, as a generalized tradition, circumcision is regarded as a procedure through which one enhances his body (in terms of height and fitness); penis (in terms of form and size); sexual and reproductive capacities (to be able to cause pregnancy); and relational opportunities (to enable them to court and marry a girl, or be preferred by women). The campaign needs to develop anti-circumcision strategies based on these beliefs. For instance, the advocacy may communicate about the lack of connection between circumcision on the one hand and physical growth and procreative capacity on the other. In this particular argument, it may be useful to cite countries where, despite the absence of circumcision, their men are fathers, tall and robust. The overall focus should be modification of the broad belief that circumcision equals manhood or vice versa — that it is a necessary procedure to become 'men' or that to be 'men' one should experience it. Efforts to rectifying masculinity concepts are not new in the Philippines. In the contexts of family planning[7] and domestic violence,[8] some Filipino men's masculinity beliefs are likewise being changed. The anti-circumcision campaign's work on masculinity, which is key to its cultural grounding, therefore, forms part of the broad advocacies to modify problematic masculinity concepts. As in other campaigns, the anti-circumcision effort should explore alternative role standards for Filipino masculinity.

Masculinity-belief modification is going to be a difficult process for the Philippine-based anti-circumcision advocacy, particularly that the specific concepts of being 'men' — in the context of its association with the procedure — are socialized to prepubescent boys. In their young ages, pre-teen individuals are most certain to conform because it is social acceptance, not rejection, that they so desire. Without conforming, these boys would be publicly ridiculed as *supot* (uncircumcised) or *tiktikon* (someone with smegma in his penis). In fact, echoing the voices of circumcised Filipino males and circumcisers themselves, CIRCS showed that peer-group teasing constituted a central explanation for wanting to be circumcised. Drawing from my own ritualistic circumcision experience, which was performed by a medical doctor in a public facility, I would say that many of us Filipino males tend to get circumcision immediately upon 'coming of age,' primarily in response to peer teasing and pressure and, oftentimes, this occurs without or with just little understanding of the essence of the tradition. Because of our peers and our strong desire to be socially accepted, we endure the procedure's attendant physical pain, specifically from the non-anesthetic circumcision performed in a traditional manner.

Notwithstanding the potential difficulty of reversing peer pressure, it is without doubt that the anti-circumcision advocacy — for it to become effective — needs to place the peer group at the heart of its campaign. In particular, it has to address this group so that its teasing and ridiculing of teenage boys foregoing circumcision are minimized. Again, modifying peer groups' manhood concepts is relevant for use in this respect. However, it is important that parallel efforts should be put in place alongside the belief modification approach. For example, a community-based support group consisting of parents, teachers, women, and male peers that will explain to teased boys the non-necessity of circumcision, and also to discourage and reprimand the other boys' teasing behavior. If the CIRCS data, underlining that 11.4% of the circumcised respondents believed that the procedure had not had any effect on them, were any indication, there is a number of Filipino males who appeared to be unconvinced of the necessity of the tradition — they could be prospective supporters and should be tapped for the campaign. As well, it is not farfetched that some parents, teachers, and women may also be ambivalent about circumcision and they can similarly be encouraged to participate. The pivotal approach is to organize support groups towards forming critical masses of campaign advocates. Definitely, for parents, teachers, and women to join these groups, they have to be educated, as the Filipino males should be, on circumcision's non-necessity vis-à-vis the pursuit of masculinity, and on the procedure's attendant pain and complications. The campaign's current message, which highlights the pain resulting from the procedure, is evident in CIRCS data and, therefore, should be continued and further enhanced, but subsequently, for additional effect, it should include evidence and testimonies on circumcision-related complications.

The campaign's work will be incomplete without strategies that will seek to convince circumcisers (medical and traditional providers alike) on the lack of a necessary justification for the Philippine circumcision and, therefore, on the need for them to stop providing it. The strategy will certainly be far from being straightforward due to two realities. One, male circumcision is not listed officially as a medical procedure in the country, and two, traditional circumcisers, whose roles are systematically replaced by newer batches of providers as shown in the CIRCS data, are not part of the modern medical health system (in fact they do not need a license to perform the procedure). The development of approach geared towards seeking the campaign involvement of the circumcisers would be particularly complex and it would require careful and serious thought. It is worth knowing, though, in the case of the circumcisers interviewed in CIRCS, that many of them were guided by some judicious standards in the way they carried out the procedure — for instance, they considered a number of factors before circumcising their clients; at least, some were cognizant of the need for making the practice safe. Regarding safety, the campaign (or another initiative) needs to attend to this issue in immediate terms, so that, while the elimination of circumcision is being worked out, those boys who are still undergoing the procedure are safeguarded from harm.

As could be seen, the Philippines' anti-circumcision campaign needs to harness the engagement of several strategic groups, and, as a matter of fact, of the whole community itself, to be able to advance its culturally grounded work. Since the work is largely social in focus and substance, the campaign will benefit if it also involves the expertise of social scientists (presently, the anti-circumcision advocacy is spearheaded by a group of medical doctors).

Acknowledgments I thank Loyd Norella, Bruce Ragas, Redentor Rola, Michael Sibbaluca and Christian Tena for assisting in the conduct of interviews. The Ford Foundation/Indonesia through Dr. Meiwita Budiharsana and the Australian National University Department of Demography through Prof. Terence Hull are cited for supporting the study on which this report is based.

References

1. Joson, R. (2004) Patient care: Should I have my son circumcised this summer? Pinoy MD, the Website for Filipino Doctors, accessed 14 June 2004, available at: http://www.pinoy.md/modules/news/article.php?storyid579.
2. Hull, T. and Budiharsana, M. (2002) Male circumcision and penis enhancement in Southeast Asia: Matters of pain and pleasure. Reproductive Health Matters, 9, 60–67.
3. Morales, D. N. and Monan, J. (1979) A Primer on the Negritos of the Philippines. Manila: Philippine Business for Social Progress.
4. Maab, G. (1997) Tule: Cutting memoirs. Parents. May, pp. 19–20.
5. Circlist (1999) Is circumcision during youth embarrassing? Accessed 14 June 2004, available at: http:// www.circlist.com/rites/shame.html.
6. Lee, R.B. (2006). Filipino experience of ritual male circumcision: Knowledge and insights for anti-circumcision advocacy. Culture, Health and Sexuality, May–June 2006; 8(3): 225–234.
7. Clark, S., Flavier, J., Jimenez, P., Lee, R. and Solomon, H. (2005). Assessment of Men's Involvement in the Philippines. Manila: United States Agency for International Development.
8. Lee, R.B. (2002). Filipino Men and Domestic Violence. Manila: Social Development Research Center, De La Salle University-Manila.

Traditional Male Circumcision in West Timor, Indonesia

Practices, Myths, and Their Impact on the Spread of HIV and Gender Relation

Primus Lake

Abstract A complete description of harmful practices of male circumcision of both Atoni Meto and Belunese in Timor is presented, based on my research, Plan Indonesia, and Indonesia HIV/AIDS Prevention and Care Project (AusAID), funded by the Ford Foundation. After my first research on Traditional Circumcision of Atoni Meto in 1997, I worked with Plan Indonesia to campaign a healthy circumcision program in 33 Atoni Meto villages. In 1999 and 2004, I did research on traditional circumcision of Belunese people, also in Timor. Based on the result of my research, from 1995 up to the present, I am working with IHPCP (AusAID), promoting healthy circumcision and sexual health to the people of 101 villages in Belu Regency on the border of the Democratic Republic of Timor Leste.

The majority of people in West Timor, Indonesia, are the native, Atoni Pah Meto, hence referred to as Atoni. The number residing in the region is 803,394 or 61% of the population (2002). These indigenous people are Christian (both Catholic and Protestant) with low education, and depend upon dry land agriculture. Up to present, the Atoni Meto has extensively practiced the traditional circumcision heritage from their ancestors.

The following is a brief description of the traditional circumcision, including who undergoes circumcision, when, where, and how circumcision is best performed, and compulsory post-circumcision copulation.

Male Atoni Meto can only be circumcised when they have come of age (generally between 16 and 45 years old). Commonly, they are circumcised before they get married; however, many also are circumcised after marriage and having had children. There are some reasons why a male is circumcised when he comes of age. It is widely believed that, if a person is circumcised in his childhood, the foreskin of his penis can sprout up again. If a person is circumcised in the early age, his body will be stunted, though it may get bigger. The foreskin of the penis of an infant is still closed, since he has never experienced sexual intercourse, so technically it causes a problem for the circumciser in preparing the clips/pin and pulling out the foreskin. That is why the circumciser will only circumcise males who have had

Faculty of Social and Political Sciences, Nusa Cendana University, Kupang, Indonesia sexual intercourse. It is impossible for an infant who has not had intercourse to carry out post-circumcision copulation, especially the so-called "sifon" (the first sexual intercourse after circumcision). Since "sifon" is impossible, it is feared that the infant may experience sexual impotency.

Circumcision does not occur at a random time. In the rainy season, circumcision takes place when corn is in bloom, young corn is in season, or when rice turns yellow (from January to April). In the dry season, circumcision takes place when "dadap" (a kind of tree) begin to blossom (July to September). It is believed that the bodies of those circumcised in these seasons will thrive and their faces will look fresh and reddish, reflecting a healthy condition.

Most circumcisers choose a hidden pond in the river as the place for circumcision. The pond is useful for patients in that they can soak themselves in the water before and after circumcision. The purpose of submerging before circumcision is to get the body cool in order to avoid hemorrhage. Besides, the foreskin becomes soft, so that it is easy to pin and cut. The purpose of submerging after circumcision is to wash blood from the body. The tools used to circumcise are usually a knife blade (or razor blade) and pair of clips. These tools are not sterilized, and the same tools are used to circumcise a group of people all at once.

A ritual ceremony related to the process of circumcision is called "nain fatu" (literally, counting pebbles). Before cutting the foreskin, all the patients are asked by the circumciser to count pebbles taken from the bank of the pond. (Every pebble taken by a patient represents one woman the patient has had sexual intercourse with.) Thus, the number of pebbles taken shows the number of female copulatory partners (excluding the wife, if the patient is married). A ceremony is actually performed by the circumciser for the pebbles. This is a rite for purifying so that the circumcision is successful. Since every patient has "pebbles of sin," the ritual ceremony for counting pebbles is an indicator of the attitude about promiscuity.

Most circumcisers discard the upper part of the foreskin and the side part of penis cap, while the lower part is not discarded and, therefore, this part becomes swollen, which they call "kaulili" because its shape is similar to a ripe tomato.

The process of circumcision is as follows: all the patients are asked to soak themselves while gently pressing their penis. When they get cold, a clip is fixed on the penis by pulling the foreskin to the back, and the foreskin of the upper part of penis cup is pinned using clips until it becomes flat. Next, the clips are tied up and the foreskin is cut. After the circumcision, the patients soak themselves in the water again, while pressing gently on the penis until bleeding stops. Finally, the circumciser treats the patients in the traditional way (using traditional ingredients) or by using modern medicines.

One of the unique aspects in the traditional circumcision of Atoni Meto is an obligation to have sexual intercourse with three women after circumcision as a part of the circumcision ceremony. The Atoni Meto recognizes three types of sexual intercourse after circumcision, namely 'sifon,' 'saeb aof,' and 'ha'ekit.'

'Sifon' (or sufun, maputu, or tape kaulili) is the first post-circumcision copulation, which is compulsorily committed by patients when the wound has not recovered yet. According to the Atoni Meto, the action of circumcising causes the penis tissues to be cut off, and the blood in the body comes out. In addition, the knife blade used for circumcision causes a 'heat' on the penis. If the heat is not discarded, the penis will not function properly, and it can cause sexual impotency. Just like the heat of fire can weaken the iron, and the cool of water can strengthen it again when gilded; likewise, the weak penis due to circumcision will get strong again after gilding (sifon) in the vagina. This is why the first sexual intercourse is called 'sifon,' which may mean 'gilding.' 'Sifon' is intended to discard the heat, which is also called 'polen maputu' (discarding heat).

As previously cited, 'sifon' is to be performed before the wound recovers. It aims at discarding the heat in the patient's body out through the wound, and liquid from the vagina can penetrate into the wound so that the wound can recover at once, and sexual lust may increase. That is why a male person committing 'sifon' will not clean his penis during the day so that the liquid from the vagina can penetrate the wound.

When is 'sifon' to be performed? Most circumcised patients had intercourse on the seventh day after circumcision. At the time the 'sifon' takes place, wounds are still big and the 'kaulili' (the swelling under the penis cap) is getting ripe. When 'sifon' occurs, the 'kaulili' is broken in the vagina. It is really a fact that in some places, one of the aims of 'sifon' is to break 'kaulili.' Therefore, 'sifon' is also called 'tape kaulili,' which means 'breaking kaulili.'

A female partner in 'sifon' should be known to be accustomed to sexual intercourse because a female partner in such criterion is assumed to have a large vagina. This makes penetration by the wounded and swollen penis easy. It is such an important reason that many circumcised patients commit 'sifon' with widows or prostitutes. A female partner for 'sifon' is usually sought by a circumcised patient himself but sometimes is arranged by the circumciser. Nowadays, most circumcised patients choose sex workers (prostitutes) in the region as their partner for 'sifon.'

The second sexual intercourse after circumcision is 'saeb aof' (or saeb na, oe'kane, poak bet, haumeni, manikin). The aim of 'saeb aof' is to get the body fresh (recovery) from bleeding after circumcision, and to eliminate the rough tissue around the circumcision scar. Generally, the second copulation is committed only when the wound has healed. The female partner for this second copulation is usually a younger female, sought by the patient himself.

The third sexual intercourse is called 'ha'ekit' (taknino, hainikit, hauhena). The purpose of this is to clean the penis, so that it looks smooth and shiny. The female sex partner is usually younger than the first two and sought by the patient himself.

Circumcised patients who have wives or permanent partners may not copulate with them, respectively, in the three kinds of post-circumcision copulation, due to a belief that such copulations may cause sexual diseases (tnan menas).

According to Atoni Meto, there are some essential reasons that encourage males to have themselves circumcised. The foreskin contains dirt (smegma). When having sexual intercourse, the smegma can be left in the vagina, which may cause the female partner to suffer from sexual diseases. The body of an intact man smells like an uncastrated male goat. Children born from an uncircumcised father often get sick. An uncircumcised man will never satisfy his sexual partner because of premature ejaculation. This may embarrass him because he is not regarded as a true man.

Possibly, in the long run, his partner (his wife) may have a love affair with some one else. An uncircumcised man is usually mocked by villagers. Such embarrassment forces him to let himself to be circumcised, even though he does not want to do so. An uncircumcised man will get older quicker than a circumcised one. A male should be circumcised because it is required by tradition. The above-mentioned myths have led most males of Atoni Meto, including educated persons, to wish to have themselves circumcised, according to this tradition.

Some aspects in the traditional circumcision of Atoni Meto need to be addressed because they are in contradiction with religious, moral, gender perspective, human rights, and reproductive and sexual health.

Post-circumcision copulations are against religious values because a female person (not wife) becomes the sexual partner of a circumcised patient in the copulation. In the light of religion, such is an action of committing adultery.

In addition, sexual intercourse following circumcision tends to subordinate the prestige and value of the female, since the females become the object of sexual interest of males. In 'sifon,' a female's vagina is treated like a rubbish basket for discarding the 'heat,' and, in the next copulations, the vagina is used as a means to clean a male's penis. This clearly shows an act of violence toward women.

From a health point of view, the traditional circumcision of Atoni Meto potentially has its role as the medium of sexually transmitted infections and HIV, due to the sharing of unsterilized tools used in circumcision and obligation of post-circumcision copulation with females who are not the permanent sexual partners.

The only solutions to overcoming various problems related to the deadly traditional circumcision of Atoni Meto is either by carrying out safe circumcision or by promoting genital integrity.

Circumcision: If It Isn't Ethical, Can It Be Spiritual?

Miriam Pollack

Abstract For the baby, circumcision is unquestionably a profound physical and psychological wound, and once recognized, is also an agonizing realization for the mother. Against our best intentions to protect our precious newborns, our culture tells us it is holy to cut, trivializing the trauma and denying the permanent damage to our baby boys' sexuality. Circumcision disempowers the mother at the height of her deepest biological impulse to protect her newborn. In trust and ignorance, we surrender to the authorities of tradition, to the pressures of family and to the prevailing myths of the general culture. But, the paradigm is shifting: more and more Jewish women, both in the U.S. and in Israel, are understanding that defining what is sacred must be anchored in the deepest, most abiding, and ancient of Jewish values, which is and has always been the primordial maternal passion: reverence for life. We will examine how circumcision has functioned in perpetuating Jewish identity, versus some of the deeper forces that may have influenced its adaptation by Judaism and unquestioning adherence by Jews throughout the millennia. Changing Jewish consciousness about circumcision both in the United States and in Israel has become a women's issue as well as men's. With our deepest feminine knowing, we can redefine the sacred so that future generations of Jewish men and Jewish women may celebrate their Jewish identity without the physical, emotional, and spiritual wounding associated with circumcision.

I am writing as a Jewish woman in opposition to the millennial-honored, primary Jewish ritual and *mitzvah* of circumcision. I also stand here fully cognizant of the precariousness of my position. Circumcision has been at the nexus of the centuries-long anti-Semitic argument used to "prove" the inferior, primitive, and unenlightened character of the ancient legacy of Jewish spirituality. Despite my best intentions, I am aware that my words may be used to further this perverse and widespread mentality.

For my fellow Jews who may hear or read these words, I understand how threatening they may sound, not only because of the way others have used this critique in the past and may use it in the future, but also because of the identity trauma such

---

Educator in private practice, Boulder, Colorado, USA

G. Denniston et al. (eds.), *Circumcision and Human Rights*,

a position seems to pose. Although my message is painful, my intention is to heal, not to hurt. I have not arrived at my position as one of many shame-based Jews, illiterate in my culture and tradition, with a need to distance myself from my identity. Quite to the contrary, Judaism has been an identity that I was both born into and have actively and joyously embraced. Judaism has been the home of my body, my heart, my mind, and my spirit. It has been the fulcrum of my sense of community, my imprint about the workings of the heart, the scaffolding for the contours of my mind, and the inspiration and anchor of my spirit. It has defined and illuminated my understanding of all that is sacred. The Hebrew language resonates deeply in my DNA and feels at home in my ears. The tragedy and beauty of the people and land of Israel pulse through my mind whether I am awake or asleep. I am very much a daughter of this tribe.

So, when my beautiful baby boys were born in 1978 and again in 1982, of course, without question or hesitation, I arranged for their ritual circumcisions. The foreskin, I was told, was simply a little flap of extra skin, connected with "scar tissue" to the head of the penis. The event would be over quickly, and, besides, they would not remember it.

I was totally unprepared for what followed. With my whole mother's body, I witnessed their screaming agony, screams that will remain embedded in my bones until I take my last breath. Inarticulate with shock, unable to begin to voice my confusion about how a tradition so steeped and detailed in defining and regulating the meaning of justice and mercy, could mandate this practice, I wrestled in silence for the next nine years.

With numerous injunctions, Judaism is emphatic and specific regarding the sanctity of life. The principal of *pikuah nefesh* mandates that any commandment may be broken to save a single life. The biblical and Talmudic messages are clear: we are stewards of this earth and it is our task not only to protect it and the life it contains, but more than that, *l'havdeel bain kodesh v'chol*, to make distinctions between the holy and the profane, so that we may consciously and continuously sanctify life.[1] This is the essence of Jewish teaching.

When a flyer for the Second International Symposium on Circumcision landed in my mailbox nearly a decade later, I summoned all of my strength and phoned the contact person listed. That was the beginning of an extraordinary journey, with my mentor and, now, dear friend, Marilyn Milos. For three-and-a-half days I sat through presentations on the anatomy and physiology of the penis and the foreskin, the short- and long-term damage of circumcision, as well as the psychological implications, and I wept. Deep in the darkest chambers of my mother's heart, I had sensed the presence of this shadow. How did circumcision originate and what functions does this ritual really serve?

Circumcision did not originate with the Jewish people. Most likely, its origins are African.[2] Nevertheless, the Bible makes the centrality of this ritual clear in the injunction given in Genesis 17:9–14 in which G-d commands Abraham to circumcise all the men of his household as a sign of G-d's covenantal relationship with the Jewish people. This commandment binds male Jews to G-d, and to all male Jews past, present, and future. This practice has gone on for millennia and is normative in Judaism.

And yet, we know the circumcision that was practiced in biblical times was substantially different than what is done today. Circumcision in the days of Abraham, Moses and, even King David, did not involve removal of the entire foreskin, but rather the portion that extended beyond the glans. Only in response to the threat of assimilation during the Hellenistic period did the rabbis instigate the requirement of radical circumcision or *periah*.[3]

As time went on, circumcision became more than a mark of a religious commandment and tribal identity. It became laminated to our psyches through the threat of continuous annihilation. Throughout the millennia, various oppressors forbade the practice of circumcision. Experienced as a frontal attack on Jewish survival, the Jewish response, understandably, has been absolute defiance, often to the point of martyrdom. To most Jews, circumcision is the sine qua non of Jewish existence. We may question G-d, but to question circumcision is tantamount to tampering with our very survival.

Although this position may make sense from the vantage point of the cyclic trauma of Jewish history, it becomes highly problematic when viewed from a more rational perspective. First of all, we might ask, how has circumcision served our survival interests, when we know how terribly vulnerable our male children and men have been during times of persecution. How many tens of thousands of Jewish babies, children, teenagers, and adults have we lost when all that the oppressor needed to do was pull down pants to expose the undeniable mark of Jewish identity? A second consideration is the status of our Jewish men in our contemporary, more secure environment here in the United States. Here, we have thousands of circumcised Jewish males who are Judaically ignorant and unaffiliated. Has altering their penises effectively conveyed to them their spiritual heritage or secured their group identity? In the Middle East, is a Jewish penis distinguishable from a Muslim? Finally, the assumption that circumcision secures Jewish continuity is predicated on the assumption that our survival is primarily male dependent. It is true that circumcision confers special status on the Jewish male. Yet, numerous Talmudic and rabbinic treatises are explicit concerning the sacred role of the mother in the raising of Jewish children and the creation of a Jewish home. Traditionally, the mother is seen as creating the container for nurturing the transmission of the Jewish heritage. Cutting exquisitely sensitive sexual tissue ensures neither the physical survival of our people nor the perpetuation of our spiritual legacy.

Nevertheless, crushing and slicing a baby's foreskin does ensure both short- and long-term damage to this male's sexuality and well-being. As is true in all other aspects of biology, altering form invariably alters function. The numerous and profound physiological consequences of circumcision have been well documented elsewhere in this book. The magnitude and significance of the baby's suffering also deserves our attention. The piercing shrieks, flailing head, and tremulous chin, or worse, the dissociative silence of babies undergoing circumcision has been insufficient for many to acknowledge the extreme degree of pain experienced by these male infants. We now have data, replicated by numerous scientific studies, which have measured and quantified babies' suffering by analysis of cortisol levels, respiratory and heart rates, oxygenation levels, duration and pitch of cry, as well as facial expression. The findings are indisputable: babies undergoing circumcision, even with the administration of pain-blunting techniques, experience quantifiably extraordinary pain. According to Dr. Robert Van Howe, in his chapter in Ref. 4 found "neonatal circumcision … produces long-term alterations in neurological response to painful stimuli."

The inextricable nature of mind and body has been well established: we can no longer deny that trauma of this magnitude can possibly be psychologically insignificant. Quite to the contrary, circumcision has been shown to disrupt the establishment of breastfeeding, an elemental part of the maternal-infant bond.[5] On an even deeper level, it may fracture or shatter an infant's trust in his mother. In utero, a baby is imprinted to the mother. She is his universe, source of all nurturance and protection. When that protection is violently breached, the baby's entire sense of safety in that primal relationship is invariably compromised, creating a deep fissure in the most foundational developmental task of all humans: the establishment of trust.[6] Upon this foundation, as Erik Erikson contended, rests all subsequent development. The baby boy's sense of wholeness is broken.[7] Would it be difficult to understand how this might affect his future level of comfort with women, with his own sexuality, or any level of intimacy?

Circumcision affects mothers, as well. Raw from the miracle of her birth, a woman is at the peak of her primary mammalian instinct to nurture and protect her newborn child. It is precisely at this moment, or eight days later, that her tradition or culture confronts her with the "necessity" of surrendering her beautiful, tender baby to the men so that his exquisitely sensitive male organ may be crushed and cut to bond him with his people and a male imaged G-d. Though silent and complicitous, she is often catapulted into a profound conflict between her tradition and her entire life-giving feminine biology. When a mother is admonished to distrust her most elemental instinct to protect her newborn infant, what feelings can she ever trust? In this way, circumcision contributes to the repatterning of the power dynamic in the family, creating a hierarchical structure with the mother and baby subordinate to the father. Her maternal authority has been subverted, her bond with her child distorted from the beginning.

Beyond the religious commandment and the arguments of identity are even deeper issues of gender and power. Many would argue that circumcision arose to compete with the strong matrilineal principal that has characterized Judaism for thousands of years. Circumcision permitted the transfer of inheritance through patrilineal descent and conferred privileged status to men.[8] However, according to *halacha*, the body of Jewish law, as well as the contemporary Israeli Supreme Court, a Jew is one who's mother is Jewish. Maternal lineage trumps circumcision, even today.

The issues of gender and power do not confine themselves to tribal and familial arenas. They reach into the most intimate areas of our lives. Circumcision affects sexuality, which directly impacts our most significant relationships. Since the glans of a circumcised penis thickens and becomes desensitized as a man ages, imagine the extra pressure on a couple as the woman approaches menopause with decreasing vaginal lubrication. Another scenario presented itself to me when I shared some of my work with a male friend of mine who is gay. "Oh, that's interesting, "he commented when learning about the consequences of circumcision, "I hardly have any sensation in my penis. I find my anus far more sensitive. If I hadn't been circumcised, I wonder if I would have made different choices about how I have pursued sexual pleasure."

"... Just a little flap of skin ..." Did most rabbis understand the sexual function of the foreskin? Probably not. But some did. The great Rambam, Moses Maimonides, the twelfth-century philosopher, physician, and Talmudic scholar, did. Here is what he had to say in his famous book, *The Guide of the Perplexed*, written in 1160:

> The fact that circumcision weakens the faculty of sexual excitement and sometimes perhaps diminishes the pleasure is indubitable. For if at birth this member has been made to bleed and has had its covering taken away from it, it must indubitably be weakened. The Sages, may their memory be blessed, have explicitly stated: It is hard for a woman with whom an uncircumcised man has had sexual intercourse to separate from him (Genesis Rabbah LXXX). In my opinion this is the strongest of reasons for circumcision. (P.609)[9]

There they are, the twin fears: the fear of woman and the fear of pleasure. Circumcision is the antidote, which both assuages and perpetuates these ancient terrors. This is the achievement and true purpose of circumcision. It achieves this by violently breaching the maternal-infant bond shortly after birth, by mutilating and marking the baby's sexual organ, by disempowering, "taming," the mother at the height of her instinctual need to protect her infant, by bonding the baby to the men and the male-imaged G-d, and by psychosexually wounding the manhood still asleep in the unsuspecting baby boy. Circumcision may be an ancient rite, but it is wrong. It is wrong in terms of Jewish values, for it violates the most fundamental Jewish principles of sanctifying life. Taking a knife to an infant's genitals, creating shocking pain, and permanently diminishing his capacity for sexual pleasure cannot be a *mitzvah*.

Repeatedly, history has shown us the remarkable ability of Judaism to mutate in practice and retain the integrity of its spiritual legacy. Judaism was not vanquished when the first temple was destroyed, nor when the second temple was razed. The discarding of animal sacrifice as the primary mode of worship did not result in an unraveling of Jewish spirituality or continuity. Neither will the cessation of genital cutting result in a diminution of the physical continuity or spiritual transmission of the Jewish heritage. In Deuteronomy, Chapter 30, verse 19, the Torah commands us as follows:

> I call heaven and earth to witness before you this day, that I set before you life and death, the blessing and the curse; therefore choose life, that you may live.[10]

To choose life: this is our core commandment. As Jewish women, both here and in Israel, connected with the depths of our maternal wisdom, we are beginning to raise our voices to name what is sacred. With a *bris b'lee milah*, a covenant without circumcision, we can choose life, and choose Judaism, for us and for our beautiful, perfect babies.

L'hayyim!

References

1. Talmud Tractate: Yoma 85a; Tractate: Hullin 10a.
2. DeMeo, J. The Geography of Male and Female Genital Mutilations. Sexual Mutilations: A Human Tragedy, eds. George C. Denniston and Marilyn F. Milos. New York and London: Plenum, 1997:4.
3. Kohler, K. Circumcision. The Jewish Encyclopedia. New York: KTAV Publishing House, Inc., 1964:93.
4. Van Howe, R. S. Anesthesia for Circumcision: A Review of the Literature. Male and Female Circumcision: Medical, Legal and Ethical Considerations in Pediatric Practice, eds. George C. Denniston, Marilyn F. Milos. New York: Kluwer Academic/Plenum, 1999:90.
5. Dixon, S., Snyder, J., Holve, R., Bromberger, P. Behavioral effects of circumcision with and without anesthesia. J Dev Behav Peds 1984;5:246–250.
6. Laibow, R. Circumcision and its relationship to attachment impairment. Paper presented at the Second International Symposium on Circumcision. San Francisco, CA. May, 1991.
7. Erikson, E. Childhood and Society. New York City, NY: Norton & Co., Inc., 1950, 1963:249.
8. Teubal, S. Sarah the Priestess: The First Matriarch of Genesis. Athens, OH: Swallow Press/ Ohio University Press, 1984:66.
9. Maimonides, M. The Guide of the Perplexed. Trans. Pines, S. Chicago and London: University of Chicago Press, 1190/1963: Vol. 2, Part 3, 609.
10. The Holy Scriptures. Philadelphia: Jewish Publication Society of America: 1955: Deuteronomy 30:19.

Garbage In, Garbage Out

Coding, Reporting, and Analyzing Circumcision Data in the United States

Georganne Chapin

Abstract Calculating the prevalence and rate of routine (i.e., non-medically necessary) circumcision in infants and children requires an understanding of hospital and physician billing and coding (including such practices as bundling and DRGs), as well as medical practice conventions. These appear to vary depending on locale, insurance coverage (or lack thereof), coding and billing competence, and the level of sophistication and accuracy of data recording, reporting, and analysis in a given entity—be it hospital, insurance company, or government agency. Using data from New York's Medicaid program, this paper will demonstrate the difficulty of collecting valid service and cost data for routine circumcision. It will also show how both states and insurers consistently under-report both the frequency and cost of circumcision. Finally, it will suggest some implications of this under-reporting for advocates working to end non-therapeutic circumcision in the United States.

Those of us who want to see the abandonment of routine circumcision — the removal of normal genital tissue from a healthy child — pay close attention to statistics and trends showing, for example, the number of parents who say they would refuse or agree to circumcision of their newborn infant boy, the number of states refusing to fund the procedure through their Medicaid systems, and the statistics regarding the incidence of circumcision by state or nationwide. While we have no trouble accepting that, in the first example, Internet surveys are unreliable because of the manner in which they are disseminated and their ability to be manipulated by electronic "voting" tricks, we are less likely to question the reliability of "official" statistics, such as national data compiled from hospital discharge summaries or state-generated reports on Medicaid-funded circumcisions. It is precisely these latter data sources that activists cite to prove the falling rate of routine circumcision in the United States, but it is this author's opinion that such data are grossly inaccurate, and that their inaccuracy leads to a significant under-estimate of both the number of circumcisions being performed in the United States and the cost of those circumcisions.

President and CEO of Hudson Health Plan and faculty member, Pace University School of Law, Hudson Health Plan, Tarrytown, New York, USA

It is no secret that the United States healthcare system is both expensive and confusing. We spend in multiples per capita what other modern nations spend, and what we get for it is significantly worse health — more morbidity, higher mortality rates among most sectors of the population — than that of people in other countries with comparable life styles, income and educational levels.[1] There are a number of reasons that healthcare costs more per capita in the United States. One is pricing — we pay more for drugs, for example, than people in any other country and our physicians earn higher salaries than those in other countries. Another reason is that Americans get a lot of duplicative services, unnecessary treatments, and interventions. A third is the administrative cost burden that results from the fragmentation, duplication, and general chaos that reigns every day in physician and other provider offices, hospitals, pharmacies, government, and insurance companies — in short, among all parties who are selling, delivering, billing for, or paying for services.

Within each of these cost-drivers, reasons for waste vary. One huge problem is the irrationality and inequities that result from a fragmented coverage system where some individuals are entitled to a lot of services, others are entitled to no services, and people flip-flop back and forth between these categories, depending on what (if any) insurance coverage they have at any given time. Not surprisingly, in a market-driven healthcare economy, every incentive exists for providers to maximize the opportunities for revenue by delivering only those services that are reimbursable, and by coding and billing for services in such a way that will maximize the probability of obtaining payment.

Medical coding is a field of professional expertise unto itself, and it is beyond the scope of this paper (or the author's expertise) to delve into its complexities. What I will do instead is illustrate — with data from one small company that pays medical claims to doctors and hospitals — the unreliability of claims as sources of circumcision diagnoses and procedures, contradictions and confusion in the data, and the consequent underestimate of the number of infant circumcisions that actually take place. The examples I will give come from one HMO (managed-care organization) located in New York State and serving Medicaid recipients in six counties under a contract with the New York State Department of Health. The author has been the chief executive officer of this managed-care organization (MCO) since 1989.

Medicaid was created in 1965,[2] as a way of funding Medical services for extremely low-income people. While the federal government pays a portion (usually about half) of Medicaid costs and establishes requirements for service delivery, quality, and funding, each of the states administers Medicaid separately. Within certain parameters, states are also given discretion over income requirements for eligibility and services that the program will fund.

Payments for services under Medicaid were originally based on the prevailing (at the time) "fee-for-service" model. Under fee-for-service, the doctor or hospital that provided a covered visit or service to a Medicaid patient would simply send a bill to the Medicaid agency in the state where the patient resided, and the state would issue a payment to the provider for the amount allowed by that state's Medicaid fee scale. Over the past two decades, however, in an effort to contain costs and improve access, states have adopted a managed-care model for their Medicaid programs, and many Medicaid recipients have been transferred from the "fee-for-service" system to private (including for-profit and not-for-profit) health maintenance organizations. In 2005, more than half of all Medicaid recipients (63%) nationwide were enrolled in managed-care programs of some type. In New York, approximately one-half of all births are funded by Medicaid (Medicaid births in 2001 ranged from 19.7% in New Hampshire to 56.2% in Louisiana, while the number nationwide exceeds 40%).[3]

Data from Medicaid managed care provides a window into the prevailing practices and problems related to coding and billing for circumcisions nationwide. Although the statistics discussed here were submitted to just one payer organization, it is important to remember that the claims from which they are drawn originate at the provider (i.e., physician and hospital) level and reflect the prevailing coding and billing practices in the current healthcare environment. Further, because the same diagnosis and procedure classification systems are used regardless of payment or funding source, conclusions regarding systematic underestimation of circumcision incidence can be generalized for both Medicaid- and commercially-funded circumcisions throughout the United States. Routine infant circumcision as a medical procedure is rather unique insofar as medical procedures go. It is a *medically unnecessary* surgery, performed on a *healthy* child, in a *hospital*, often *by a practitioner whose specialty and principle role concerns attending to another patient* — the mother.* Given these unique features of infant circumcision, and the market-driven, complex, and fragmented healthcare-payment system, it is not surprising that peculiarities in billing and coding for the procedure have arisen. In 2005, HHP was billed for 1,492 live births that occurred to women who were enrolled in our public insurance programs — 702 male and 790 female. (Of the original claims submitted to the health plan, fully 13% did not indicate the newborn's sex, but this was resolved over time with the submission by the hospital of birth certificates; see Fig. 1).

Males	608
Females	690
Unknown	194
Total	1,492

Fig. 1 Live births by gender on original claims submitted to Hudson Health Plan, 2005

*Who actually performs circumcision is a rather hotly debated question; anecdotal evidence suggests that many pediatricians wish to disavow their role in promulgating the practice — a typical response to the question "Do you do circumcisions?" being "Not unless I have to," or even, "No, that's the obstetricians' job." Logically, depending on local practice patterns and the distribution and acceptance of specialists (i.e., obstetricians), family practice physicians, and nurse midwives as primary birth attendants, the type of practitioner performing circumcisions also varies by geographic area and even by hospital. In teaching hospitals, residents in obstetrics, family practice, and pediatrics may all be required to perform newborn circumcisions, bearing little or no relation to whether the latter two will ever do the procedure again. Similarly, a midwife practicing under supervision of an obstetrician may actually perform the surgery, although the bill will be generated by the OB.

For that same year, 233 claims submitted were associated with[1] newborn circumcision, which, if taken as a percentage of male children born into the health plan, comes to 33%. However, more than 17% (41) of those claims were originally billed under the female sex — one under the female twin, two in error, and the remaining 38 under the mother's record because the mother's insurance status was continuing, while the newborn's needed to be approved. See Fig. 2.

As for which providers were billing for circumcision, in the 2005 HHP sample, 151 (65%) claims were submitted by obstetricians, 35 (15%) were submitted by urologists, and two (under 1%) from family practice physicians. Most notably, *only one* of the 233 claims was submitted by a pediatrician. The other claims came from hospitals (with no provider specified), surgeons and (!!) even a radiologist.

While most insurers use claims-editing systems that are designed to pick up certain discrepancies (e.g., an obstetrical claim submitted for a male; a routine circumcision on an adult woman), there is absolutely no standard practice as to whether such claims will be paid (essentially, in order to avoid additional wrangling and paperwork) or rejected/pended, and the provider given an opportunity to resubmit the proper information. Either way, the number of final paid claims will inevitably be lower than that of the original billed claims, and the data from those original claims that go unpaid will be lost.

As stated above, nearly but not all circumcision-associated claims received by Hudson Health Plan were generated by physicians. Most were billed under the Routine Circumcision "V" code, V50.2 (see Fig. 3). "V" codes are described in the

Gender of patient on claim	Circumcision-associated claims
Male	192
Female	41
Number of circumcision claims	233

Fig. 2 Gender of patient on 233 original circumcision claims received by Hudson Health Plan, 2005

		Number of claims		
2005		INST	PHYS	Unique utilizing member
Age	550.92 Bilatinguinal hernia	2	0	2
	752.61 Hypospadias	0	1	1
	V30.00 Single LB in-hosp W/O CS	0	4	4
	764.2 Fet mal W/O LT-for-dates	0	1	1
	V50.2 Routine circumcision	1	199	192
	605 Redun prepuce & phimosis	3	30	31
	607.89 Disorder of penis NEC	0	2	2
	Total	6	237	233

Fig. 3 Circumcision in newborn members by diagnosis and claim type, Hudson Health Plan, 2005

[1]Not every claim associated with circumcision was actually accompanied by (or able to be matched later to) a claim for the procedure.

ICD-9-CM chapter, "Supplementary Classification of Factors Influencing Health Status and Contact with Health Services," as being designed for occasions when circumstances other than a disease or injury result in an encounter or are recorded by providers as problems or factors that influence care.[4] The other frequently used diagnostic code was # 610, "Redundant Prepuce & Phimosis"; the bogus nature of this diagnosis has been well documented elsewhere.[5]

The importance of noting the diagnostic codes used to bill newborn circumcision relates to payer policies. In New York, for example, routine infant circumcision is a covered benefit under Medicaid. Therefore, both the State and its subcontractors for the Medicaid Managed Care program[2] will pay for newborn circumcisions billed with a "V" code (as well as with any other diagnostic code, bogus or not, indicating genital pathology). Refusing to pay for V-coded procedures, however, is in itself not enough to ensure that such procedures will be discontinued, so long as alternate "pathological" diagnoses, such as "redundant prepuce and phimosis," are allowed to go unchallenged.[3]

A number of individuals and organizations have attempted to analyze the number of routine circumcisions funded under Medicaid. A paper published in 2001 by the International Coalition for Genital Integrity estimates that just over 300,000, or 25%, of all routine infant circumcisions performed in the United States were paid for by Medicaid.[6] Both the percent and number are almost certainly underestimated. With regard to the percentage, as explained above, over half of Medicaid recipients, and a higher percent of mothers and infants, are enrolled in Medicaid managed-care programs. Both provider and State practices regarding data related to those births is variable, with some parties considering only fee-for-service Medicaid as "Medicaid" and Medicaid managed-care births as "commercial insurance," and others considering such births as paid for by Medicaid.

Insofar as the number of circumcisions performed, errors and underreporting of circumcision data occur at every level of the Medicaid healthcare delivery and payment system. First, at the provider level, a significant percentage of billed claims typically have internal inconsistencies and error. Hospital charts too often omit both diagnoses and procedures. Second, where multiple providers are involved, discrepancies among claims associated with a single event (e.g., hospital

---

[2] It is to the author's great chagrin that the agency she heads must pay for routine circumcision. For this reason, Hudson Health Plan has undertaken an educational campaign, directed to expectant mothers and primary-care physicians, regarding the lack of medical indications for circumcision and the negative consequences of the procedure.

[3] A fascinating example of code-manipulation designed to guarantee federal reimbursement was cited in a (post-presentation) news piece about a federal audit that found New York State Medicaid had approved and paid an estimated $3.2 million in laboratory claims that were coded as "family planning" services (allowed by the federal government) rather than abortionrelated services (not allowed by the federal government). Crains' Health Pulse, Friday, August 10, 2007.

[4] Medicaid dollars spent on circumcisions are virtually impossible to determine, due to the obscuring of managed-care claims payment data because of proprietary rates, the uncounted but built-in additional costs the procedure adds to hospital stays, and failure to include circumcision complications in typical circumcision cost data.

admission, surgical procedure) are common. Third, at the payer level (e.g., the insurer), proprietary system edits and anti-fraud detection software and practices) will result in the rejection of questionable claims. All of these factors are equally operative in the commercial insurance sector, meaning that circumcisions performed for privately insured babies are also undercounted.

Three additional problems plague Medicaid managed-care circumcision data. The first is that claims reported to regulatory and monitoring agencies are almost always confined to those actually *paid* (i.e., not rejected because of one or more of the myriad errors or inconsistencies discussed above) by the insurer. Second, even for paid claims, details such as the dollar cost of the procedure — are often lacking due to the use of Diagnosis-Related Groups (DRGs) or other bundling or grouping software. At every step, then, both the reported number and the reported cost[4] of circumcisions decrease. What we are left with is an utterly unreliable picture of what is actually going on — how many babies are undergoing this unnecessary surgery, and how many of our limited healthcare dollars are being spent with absolutely no medical justification.

Without further research specifically designed to explore the effect of each of these factors on routine circumcision, it is difficult if not impossible to calculate the magnitude of underreporting and thus the actual incidence of routine infant circumcision in the United States. While this may be a fruitful avenue of inquiry, probably more compelling strategies include continuing to educate parents and the general public about circumcision and working toward a universal healthcare system that would eliminate inconsistencies and bureaucratic loopholes and impose strict rules against the payment for medically unnecessary procedures.

References

1. B. K. Frogner and G. F. Anderson, Multinational Comparisons of Health Systems, The Commonwealth Fund, April 2006.
2. Medicaid's authorizing legislation is Title XIX of the Social Security Act. For a concise history of Medicaid, see Medicaid: A Timeline of Key Developments, Kaiser Commission on Medicaid and the Uninsured, see http://www.kff.org/medicaid/medicaid_timeline.cfm (last accessed September 23, 2007).
3. Number of Births Financed by Medicaid, 2002. Kaiser Family Foundation, see http://www.statehealthfacts.org (last accessed September 23, 2007); Kaiser Family Foundation, State Health Facts; Total Medicaid Births by State, Statemaster.com, see http://www.statemaster.com/graph/hea_tot_med_bir-health-total-medicaid-births (last accessed September 23, 2007).
4. K. M. Kostick, From V Codes to Z Codes: Transitioning to ICD-10, American Health Information Management Association, see http://library.ahima.org/xpedio/groups/public/documents/ahima/bok1_022421.hcsp?dDocName=bok1_022421 (last accessed September 23, 2007).
5. See, for example, F. M. Hodges, The history of phimosis from antiquity to the present, in George C. Denniston et al. (eds.), Male and Female Circumcision: Medical, Legal, and Ethical Considerations in Pediatric Practice (New York: Kluwer/Plenum, 1999) 37–62.
6. A. Craig, et al., Tax Dollar Funding of Medically Unnecessary Circumcisions Through Medicaid. International Coalition for Genital Integrity (Santa Cruz, CA, March 31, 2001).

A Treatise from the Trenches: Why Are Circumcision Lawsuits So Hard to Win?

J. Steven Svoboda

Abstract Barriers of many different types make successful circumcision-related lawsuits extremely difficult to bring. Actual cases we and others have brought show that among factors impeding progress are (1) financial risks; (2) procedural difficulties; (3) misconceptions and compassion misallocation among judges, lawyers, jury members, the media, and the general public; (4) constraints unique to circumcision lawsuits that are imposed by statutes of limitation and statutes of repose; (5) need for parental participation in lawsuits; (6) problem of damages not being atrocious enough to justify litigation; and (7) the scarcity of helpful case law. Players whose roles we will be scrutinizing include clients, lawyers, judges, juries, courts and procedures, doctors, media, and fellow activists. We will discuss the many reasons why potential plaintiffs never even make it to the filing stage. We will look at why judges and juries are starting to understand that just having a foreskin is not reason enough to have a circumcision.

Introduction

There is no one answer to why circumcision lawsuits are so hard to win.[1] It is partially related to the difficulty of any sort of litigation and partially related to problems peculiar to this highly specialized area of the law. No global answer can be offered. But approaches to the answer can be glimpsed by reviewing all the many pieces of the puzzle that have to come together in litigation. I will be looking at two cases from the last few years in which I was directly representing the plaintiffs. In one case, brought by infant Dennis Pappas[2] and his parents Cyril and Maria, I represented a wrongfully circumcised infant whose parents did not give proper consent to the procedure. In the other case, brought by lead plaintiff William Anastasian and his wife Laura Anastasian, an Armenian man, identified with a culture within which circumcision is not practiced, went to a physician for a vasectomy and instead was circumcised by mistake.

Founder and Executive Director, Attorneys for the Rights of the Child

[1] Thank you to John Geisheker for his assistance with this article.

[2] All names of parties and witnesses other than foreskin-friendly doctors have been changed.

Litigation is performance art with strict, often arcane rules and with real world results. But it is also the art of convincing all the players — willingly or otherwise — to lend their various talents toward the ultimate success of your clients. The cast includes clients, lawyers with whom you are working and on the opposing side, judges, juries, courts and procedures (very real players even though they are not human), doctors (who can appear as "percipient witnesses," i.e., witnesses reporting on examinations of parties, as experts for your side, and/or as experts for the other side), and other groups — media, activists.

Before we get to the players, let's first look at why most circumcision lawsuits go nowhere, and then at why sometimes we may hope for more. I will also discuss what limited case law does exist, helpful and otherwise. Then the stage will be set to meet and learn about the various actors, group by group. We will close with what conclusions we can draw.

Reasons Why Most Circumcision Lawsuits Go Nowhere

Most circumcision lawsuits go nowhere and, indeed, most potential legal actions of any sort do not ultimately reap substantial success. In a moment, I will look at some difficulties more or less unique to activist actions. First, let's consider issues common to all lawsuits.

The barriers to bringing any sort of legal case can be daunting. Indeed, this is the very reason why so many people hire attorneys, often at substantial hourly rates — to avoid having to confront litigation themselves. Costs and expenses are substantial, and some, such as filing fees and lawyers' hourly rates, may have to be paid before a case even begins. Typical retainer agreements require clients to reimburse their lawyers for costs and expenses. In addition, lawyers are compensated either at an hourly rate or else by receiving a percentage of any settlement or court award. Often a retainer fee must be paid up front, and additional money to cover costs and expenses may also be required at the front end. If the clients cannot pay the money, whatever the retainer agreement may say, the buck stops with the lawyer, who ends up having to fund many of his or her cases.

Law is an art, not a science. Even in cases that are much more common than circumcision lawsuits, it is often hard or impossible to gauge the likelihood of success, and/or the probable size of an award or settlement if one is received. This is all the more true in activist litigation given the small amount of precedent and the widely varying results such cases have had. Rules can operate harshly and yet, paradoxically, are often indefinite enough that even the lawyers involved cannot confidently predict ultimate results.

Civil procedure is very complex. Rules vary from state to state, of course, but also from jurisdiction to jurisdiction, according to local rules and even, as discussed below, from court to court. Mistakes can forever prevent certain remedies or rights.

Now, let's consider some matters that are unique in their effects on activist litigation.

Statutes of limitations are procedural bars to bringing cases after a certain period of time has lapsed. The purposes of limitations periods include protection of judicial resources and discouraging potential plaintiffs from "sitting" on claims that, after the passage of time, may be more difficult to resolve due to the death or unavailability of potential witnesses, loss of evidence, and so on. Statutes of limitations typically do not start running until the potential plaintiff has legal capacity, i.e., has reached the age of majority.

Some states also have statutes of repose. Under the assumption that plaintiffs lacking legal capacity can have lawsuits brought on their behalf by parents, guardians, or other agents, statutes of repose require all lawsuits of certain types to be initiated within a certain number of years, even if the potential plaintiff lacks legal capacity. Naturally, a certain tension exists here between, on the one hand, the desire for finality and for lawsuits to be brought within a certain period of time and, on the other hand, concern for minors' rights. Washington, for example, has a statute of repose, while at the same time its supreme court has found the statute of repose unconstitutional and has held that it is without effect for minors due to its violating children's rights.[1] Some states adhere rigorously to statutes of limitation and repose, but interpret notice requirements liberally. Therefore, you can still bring your lawsuit if you can show that the plaintiff did not learn of his injury until recently. The limits of these doctrines vary state to state and, in most cases, have not been determined, certainly not with reference to circumcision litigation.

The uniqueness of this procedure is that, in states with rigidly interpreted statutes of repose and limitation, the possibility of relief may be eliminated by the operation of these statutes. For any other lawsuit due to injury to a child, parents or guardians will have ample opportunity to take action. Due to the nature of circumcision, often it is the boy himself who learns of the harm when he becomes sexually active and/or old enough to perform his own Internet searches and other inquiries.

Cultural factors also play a significant role. Contrary to myths about social change lawyers eradicating racism in the Old South, law tends to follow society, not to lead it. Our puritanical culture reinforces our lack of legal and societal compassion for the screaming boys. Potential clients may be suffering from bad adhesions, from scarring, from a penis that points 90° to the left or to the right. Yet, compassion may be focused on people other than the victims of this unnecessary surgery. In the Pappas case, it surprisingly may have been misdirected to the physician who performed the procedure in the first place, who happened to be very pregnant during trial. In the Anastasian case, jury sympathy may also have been directed to the elderly physician despite the numerous malpractice cases he had lost in the past and despite his other unsavory characteristics.

Circumcision rates are dropping.[2,3] As a culture, and certainly as a legal culture, we are starting to challenge our previous understanding that just having a foreskin was enough reason to have a circumcision. Legal culture is slow to transform, but change is coming, as judges, lawyers, jury members, the media, and the general public show their slowly increasing willingness to endorse the right to genital integrity for males that is so easily acknowledged for females.

Parental involvement as givers of proxy consent provides a buffer zone that the defense bar has in the past exploited. Parents often select the procedure even without the physician suggesting it and, therefore, a bad result can be cast as the regrettable but not legally punishable result of a risk willingly assumed by the parent. Unfortunate physicians can paint themselves as commercially constrained and required by their obligations to patients to perform a procedure even while they argue that it is not well-advised. Beyond that, the natural professional desire exists to make parents happy. Acceptance of activist lawsuits is hampered by social misconceptions and compassion misallocation among judges, lawyers, jury members, the media, and the general public, which in turn are related to society's acceptance of circumcision and our country's puritanical heritage.

Parental participation or cooperation is often needed to initiate a circumcision lawsuit and all too often is unavailable. An 18-year-old boy has limited abilities to buck his parents, and limited abilities to bring a case on his own without their support. One very determined young man, threatened by his parents with their withdrawal of his college support if he persevered in a lawsuit, eventually relented and cancelled his lawsuit.[4]

Sometimes as a lawyer you may feel that your whole profession is a bit twisted. Often the best thing for a case is the worst thing for a client, in that the more problems they have had as a result of an injury, the greater the relief is likely to be. The flip side of this is a sad truth: most circumcisions do not produce atrocious enough results to justify litigation. For many reasons related to all the players we will discuss below, most potential plaintiffs never even make it to the filing stage. It is not an issue of the validity of their complaints or how "good" or "bad" they may be as clients.

For example, in one case with which we are familiar, the client's penis, as a result of a botched circumcision, was buried into his abdomen. It then had to be surgically retrieved, leaving no less than 25% of it as scar tissue. Anesthesia was deemed unusable under then prevailing medical practice, so naturally the corrective procedure (not to mention the initial procedure) caused unspeakable pain to the tiny baby. All physicians would agree this fell well below the standard of care. Yet, damages probably were not high enough to justify a lawsuit. Most foreskin retraction cases fail for the same reason, sympathetic as I always am to the parents' and child's plights and complaints.

One final, serious hurdle that can only be solved by the passage of time and further legal efforts is the scarcity of helpful case law, discussed in more detail below. Overall, the limited precedent that does exist is quite positive.

Reasons Why We Sometimes Hope for More

Litigators tend to have the heart, if not the soul, of a gambler. The possibility of the big win justifies the struggle, the frustration, the time spent on the mundane, and the all too frequent losses. In both cases, we hoped for a substantial victory that would amply compensate the plaintiffs while at the same time providing ARC with a sizable warchest. Such was not to be. Harm was clear in both cases. Yet, both cases, as is common if not universal with circumcision cases, had valuation problems.

Anastasian's loss was undeniable. By his own estimation, with his foreskin, he lost 95% of the sexual sensation he previously enjoyed. Quantifying the harm would be helpful if we could do it, perhaps by testing the sensitivity in the glans penis relative to those areas still covered in skin. The circumcision directly resulted in the breakup of his marriage, and we therefore filed a claim on his wife's behalf as well. (The two maintained a friendship after filing for divorce.) The causes of action were medical malpractice, battery, breach of contract, negligent infliction of emotional distress (termed "negligence" in California), intentional infliction of emotional distress, and loss of consortium (a cause of action on behalf of his wife due to her deprivation of his family services). The case was valued at over half a million by our expert co-counsel. We thought the Glendale venue was good for us with the high Armenian population.

Dennis' loss was perhaps even greater. The circumcision needlessly caused Dennis substantial pain and suffering and has permanently scarred him. Dennis' circumcision has led to his estrangement from the community and family of his father. The practice is unheard of in both Ecuadorian and Greek cultures. When in Greece, it is a problem for him that he has been circumcised. Naked baths are taken daily.

Dennis was circumcised by a first-year medical resident, Dr. Sarah Bernstein, who was untrained in the procedure (and happened to be Jewish). Defendant Elmhurst Hospital — which is owned by New York City — engaged in systematic discrimination against Spanish-speaking mothers from whom "consents" for circumcision were fraudulently extracted without their understanding what they were signing. Maria specifically refused circumcision in writing. Maria was never given a properly translated form to read and sign. Dates and times were obviously altered. Her "consent" form was not properly translated into Spanish, was incomplete, failed to describe the procedure, and contrary to its own stated requirements, was not properly witnessed or executed. Her "consent" was obtained without her first being informed — as is legally required — of the risks, harms and alleged benefits of male circumcision. The form was signed at the top of page two, evidently without her being shown page one in any language. At no time was the potential circumcision of plaintiffs' newborn ever discussed with Dennis' primary care physicians.

This was probably the first circumcision case brought as a civil rights case. We raised federal claims based on denial of parental rights to make decisions regarding medical procedures performed on children, discrimination on grounds of race and national origin in violation of federal civil rights, as well as the more usual state law claims based on medical malpractice, lack of informed consent, gross negligence, intentional infliction of emotional distress, and parental claims for loss of society of the child. We won a procedural decision permitting us to amend the complaint, originally designed for state court, in order to proceed in federal court based on discrimination in violation of federal civil rights law.[5]

My co-counsels in this case were two litigators experienced in bringing penile tort cases. We knew that a victory would be important to the movement and could have national impact. But the judge would not allow additional victims into the case, not in their own right (as a class action) and also not even insofar as they affected Pappas' case, to show a "pattern and practice" of discrimination by the hospital.

Scarcity of Helpful Case Law

I am aware of only a handful of reported cases centrally relating to circumcision. Only a fraction of those are favorable and likely to be relevant to other lawsuits. Following is a summary of each known case addressing a circumcision on its merits. Our work is to expand the relevance of each known victory by educating the judiciary, the bar, and the public at large regarding the right to genital integrity.

Courts often search for any conceivable basis, such as a narrow decision regarding a lack of legal standing, which may allow them to avoid addressing the potentially earthshaking (and possibly politically and/or personally treacherous) merits of such cases. At least three times, in 1989, in the Adam London case,[6] then, in 1996, in the Fishbeck v. North Dakota case brought by Zenas Baer,[7] and most recently in Baer's Flatt case,[8] courts have avoided squarely addressing the legality of male circumcision by diverting the discussion into such peripheral, procedural issues as standing. In the Pappas case, as we have seen, the federal district court judge went to extraordinary, monumentally improper lengths to prevent fair consideration of Dennis' complaint regarding his wrongful circumcision.

By contrast, a preliminary, procedural decision in the Anastasian case allowed us to amend our complaint to proceed in federal court with the discrimination claim violated federal civil rights law.[9]

When a court cannot muster any valid reason to skirt the circumcision issue, peanuts may be awarded. In a lack of consent case where the medical resident failed to read the patient's record, nominal actual damages were held available, although punitive damages are unavailable in absence of gross negligence.[10] A $20,000 general damages award was upheld in a medical malpractice action over circumcision that resulted in over two weeks of hospitalization and necessitated two incisions on the penis.[11]

Occasionally a substantial award comes through. An appellate court upheld a trial court's finding that a third-year surgical resident was negligent in modifying a circumcision technique on child, resulting in burning off of child's penis, and upheld the jury's $2.75 million award. The state was held vicariously liable since the resident was employed by a state agency, was working in a state facility, and was supervised by a state medical school.[12] A trial court's $2.25 million award was reinstated for an adult circumcision that rendered sexual intercourse impossible and reduced penis in size by 4 in.[13] A trial court's $200,000 judgment for a minor was upheld where the Plastibell device used in the circumcision did not fall off after eight days and the child suffered injury.[14]

Sometimes a court employs that rarest commodity, common sense. Despite the absence of expert testimony, an appellate court held that a trial court erred in granting a directed verdict (i.e., in requiring that the jury find for the defendant) to the defendant physician where a Gomko circumcision clamp slipped when removed, lacerating the baby's penis and causing infection and a cyst. Expert testimony is not necessary, the court held, where a lack of skill is so apparent as to require only common knowledge and experience to understand it.[15]

Nor will procedural errors or delays always knock out a case. One appellate court held that even where patient delayed trying of case until more than five years after filing, genuine issues of fact exist as to skill exercised by physician and purpose of "circumcision" operation that actually cut urethra and excised two large segments of penis.[16]

At least one other circumcision case besides Pappas addresses the destruction of and inadequate maintenance of records. A hospital was held to have violated its duty to maintain and preserve records when it destroyed records pertaining to a circumcised boy who suffered infection, a subsequent seizure and permanent, severe brain damage and disability. Its destruction ("spoliation") of evidence shifted the burden of proof to the defendants, including the hospital, to prove that the boy's injuries were not caused by their negligence. The defendants had to assume the burden of proof on both causation and medical negligence.[17]

The doctrine of res ipsa loquitur (literally, "the facts speak for themselves") was held applicable to a suit against a hospital that neglected to treat an infection, despite a black spot present on the plaintiff's penis prior to his initial release, eventually resulting in the boy losing his glans penis. Roughly speaking, the court held that the facts mean that the hospital must necessarily have been negligent.[18]

In a recent and widely publicized case, the Oregon Supreme Court ruled that a 12-year-old boy's own wishes must be investigated by the district court where a newly converted Jewish father sought to compel the circumcision of his son against the wishes of his ex-wife.[19]

On the defendants' side, summary judgment was granted to a defendant in a federal civil rights case brought under the section 1983 "color of law" principle against a prison for medical care to a prisoner, including circumcision that was not adequately explained to him and that had questionable medical value. The court held that, at most, the plaintiff stated a claim for negligence or medical malpractice for failure to inform.[20]

A triumvirate of cases basically held that in a circumcision case, you have to have expert witnesses. It was held that no action in negligence was available, despite a large scar at a circumcision site, where no evidence was presented regarding the geographic area's standard of care, no expert testimony was provided about the standard method to perform operation, and no evidence was provided showing that the physician had been unskillful in circumcising the boy.[21] Expert testimony has been held to be required on behalf of a plaintiff in an action in negligence regarding his circumcision.[22] A trial court's summary judgment for defendants was affirmed when plaintiffs failed to meet their burden of establishing standard of care and the doctor's failure to conform to it, because they did not offer required expert evidence, instead offering hearsay and the father's layman opinion.[23] Finally, no right of recovery was found where a baby is circumcised by a physician four days prior to an intended bris.[24]

Settlements usually require parties to be silent about their terms. However, we have reason to believe there may have been a settlement for over $20 million in a 1985 case from Georgia. In 1986, a Louisiana family was awarded $2.75 million by a jury after a young boy's penis was severely burned during a circumcision and had to be amputated.[25] We also know of a settlement for 800,000 UK pounds (about $1.5 million in today's dollars) for a circumcision that left an adult "grossly genitally mutilated."[26] In 2001, a Sacramento, California jury awarded $1.42 million to a boy for a botched circumcision.[27] A settlement worth $117,000 was agreed to for a circumcision done without parental permission in 1997 in the Boston area.[28] In 2003, a young man won an undisclosed settlement as compensation for his circumcision. Despite parental authorization for the procedure and its non-exceptional result, legal validity of the mother's assent was questionable as she was debilitated at the time she agreed.[29]

That constitutes every known case on the record centrally addressing circumcision. It is not much to go by, certainly not in the modern era of profligate litigation. Each case can frustratingly seem confined to its facts, yet a general principle is developing whereby courts are grudgingly starting to acknowledge at least the kernel of a right to genital integrity.

Clients

Clients naturally play a major role in the success of any case. In infant circumcision cases, often it is the parents who become the major players. Cyril Pappas loomed very large, ironically overshadowing the biggest victim, his son. Cyril was an interesting character. Born the oldest boy in a poor Greek family, he saw his only hope to improve the family's economics and joined the merchant marine after high school. At age 25, he jumped ship and made his way to the US, becoming a citizen in 1978.

Controlling Cyril in court became a big problem that we never fully solved. He was clearly emotionally damaged. Whether it was actually due to his son's circumcision was frankly unclear to us, though clearly the father thought it was, and who were we to argue with him? The possibility of physical violence against us or others could never be completely ruled out from someone who, while still on medication, had sent me no fewer than 29 letters, usually handwritten and running to 10–20 pages each, and left well over 300 voicemail messages on my answering machine, typically at three in the morning and often 10–14 messages at a time, one right after the other. We also feared a more common form of aggression, a malpractice suit. Cyril was a man who was critical of everything and everyone, a man for whom nothing was ever good enough. He mailed scathing attacks to the city's examining psychiatrist, as well as to his own, admitted imperfect medical expert. One problem for us was that given the way he played his role, instead of coming across as the victim that in fact he was, he appeared aggressive and unsympathetic.

Cyril's wife, Maria, also a fascinating individual, was a Roman Catholic from Ecuador with a Master's degree in clinical psychology. She worked as a bilingual Spanish/English teacher. Her speaking some English complicated matters, as our claim was partly based on the inadequacy of the Spanish-language consent forms she signed. Clearly, despite her education, she did not understand what she was signing. Just as clearly, she would be a good witness. We counseled her not to speak in English inside or within shouting distance of the courtroom.

Starting at age four or five, son Dennis had noticed that he was different from other Greek boys and from his father, and he was somewhat distressed by this fact. We discussed the possibility of putting him on the witness stand. His mother blocked this plan, as she felt it would signal to him that something is wrong with him. It also would be risky strategically, as it could offer an opportunity to the other side to lead him down a path toward admitting that nothing is wrong with him.

William Anastasian was also headstrong. That seems to be a trait of any client or parent whose convictions are strong enough to lead them to be willing to weather the various difficulties attendant on bringing litigation. But, he would not admit the strength of his own opinions, adopting a sort of Colombo-esque pose that he was only a client and was untrained in legal matters, while at the same time freely substituting his judgment for ours. At one point, he wanted us to try introducing a pornographic videotape as evidence of the harm caused by circumcision. Common sense would tell you what the correct response should be; this is a textbook example of evidence that would be excluded from trial as being far more prejudicial to a case than any probative value it might have. Anastasian was ostensibly an activist, though at times I wondered if this was primarily a ruse to ingratiate himself with his lawyers.

William proved his own worst enemy when, as discussed above, he refused to allow our legal expert co-counsel into the case. Ironically, William proved an unsympathetic player and not a great witness. He was too argumentative and opinionated to come across to the jury as likable, credible, and rational.

On the eve of trial, William unburdened himself to me, stating that he thought I had an "aggressive, abusive" side. He added that my approach to relationships is totally different from his, and that he has been frustrated by our relationship. Well, he is not the only one who felt that way! He was a remarkably unappreciative client. William also made some unwise decisions stemming from his headstrong personality, rejecting two settlement offers, each for $100,000, the second made the month before we went to trial. Everyone Joseph spoke to urged him to accept this offer, but he steadfastly refused it.

Lawyers

The discovery process requires each side to share with the opponents prior to trial, all evidence it intends to introduce at trial that is properly requested during discovery. Discovery entails such specific processes as oral depositions in which a witness' answers to questions from the opposing attorney are recorded by a court reporter, interrogatories or sets of written questions that the other side must answer, and requests that the other side produce all documents it intends to introduce. Discovery was initially created with the goal of simplifying litigation and making it more transparent, but of course that is not exactly how things have played out.

Perceptions of one's opponents and their likely reactions to events can play a substantial role in how a case evolves. Typically, each side tries to overwhelm and bluff the other. In the cases in which I have been involved, there have been obvious differences. In one case, our principal opponents were the State of New York and its government-employed lawyers working in a nice air-conditioned skyscraper. In the other case, we faced even fancier firm lawyers in an even nicer building. Meanwhile, we were more or less fanatics sending faxes in our pajamas. The other side had a lot of other cases and did not have any particular concern with this one above the others, whereas we had few other cases and a strong interest in protecting and defending genital integrity. Each side knew the other side fairly well. I am not sure that our status hurt us in the case, as the defendants never knew for sure what we would do next, with our seemingly unlimited resources being plowed into the case, not to mention potential assistance from other activists where appropriate.

In the Anastasian case, I tried to bring in some very skilled co-counsel with strong expertise. Before they would enter the case, they required a certain change in the fee agreement. On the one hand, changing the fee agreement midstream was arguably not the nicest thing for them to ask from the clients. On the other hand, Anastasian was very unwise to turn them down.

One co-counsel, with whom we had previously worked well, abruptly departed from his senses, showing up for a pair of depositions dressed in shorts and a Hawaiian shirt. "I'm on va-cay," he helpfully explained, acting like a player who had forgotten his lines and was improvising. Badly. He was apparently drinking in the law office's bathroom in between deposition questions. What should have been a two-hour deposition stretched out to four times that length. ARC and our co-counsel were stuck with the bill. The court stenographer, who has 25 years of experience, said the lawyer must be crazy or on drugs, as he did not ask any questions that advanced the case, and was rude at times. After the deposition, he followed the stenographer out to the elevator, asking for her opinion as to how he had done.

The hired gun on whom we eventually settled in this case, though very experienced, resembled the other side in that he was not personally committed to the issue. After doing a great job in his opening statement, he gave a lackluster closing argument. In his defense, his performance may have been impacted by confrontations Anastasian was having with him in the halls, practically firing him on the spot.

As if the quirks of your own real or ostensible "friends" are not enough, you also have to deal with the attorneys for the opposing side. In the Pappas case, they completely flouted the discovery process by withholding from us until the third day of trial a critical item we had requested, the hospital's policy manual. Yet the judge did not see fit to sanction them in any way, despite our rightful pleas that he do so. In the Anastasian case, at a mandatory arbitration hearing, at which we were supposed to try to settle our case, the opposing attorney was actually cracking jokes about our client and ridiculing him for the penile damage he had sustained!

Offers were extended shortly before trial in both cases. The Pappases received an insulting, token low-four-digit offer. I am not sure what the purpose of such offers is; surely the attorneys did not entertain the thought that the offer would actually be accepted. In the Anastasian case, the clients received a reasonable offer of $30,000, which they rejected, in a decision that seemed correct at the time. Later an offer of $100,000 was made as our opponents tempted us to avoid the bulk of discovery. The $100,000 offer was renewed the month before trial.

Judges

Each judge is different, and you ignore the differences at your peril. Local rules, sometimes set by the individual judge, render such idiosyncrasies into rules of the court. It is almost as if you were a chess player finding slight differences in the rules in each city: in Seattle, knights can also move diagonally, in Tacoma, you cannot castle on the queen's side of the board, and so on.

In the Pappas case, Judge Friedman (real name) was no less than the case's fourth judge. Judge Nickerson, the case's first judge and an excellent one, sadly passed away. The case was reassigned to a female judge, then reassigned to a visiting judge from Iowa, before finally being reassigned to Friedman. He was the case's second visiting judge, a conservative from Michigan who was given the case under the practice of the Eastern District of New York (New York City) of bringing in visiting judges to help with that district's huge caseload. He was in town for just a couple weeks and was charged with clearing up the oldest cases on the court's calendar, which included the Pappas case. Judge Friedman clearly did not like the case. Judge Nickerson had held that our case belonged in federal court, but Judge Friedman attempted to tell us, four whole years later after countless discovery conferences had been conducted, that there might be no federal jurisdiction. We were forced to write a legal brief stating why federal courts should be able to hear the case. Evidently, Judge Freidman eventually resigned himself to hearing the case and, "based on the court's own research," lifted his previous queries regarding federal jurisdiction.

Judge Friedman used his full power to orchestrate a favorable result in this case for the defendants. He telegraphed to the jurors his views of the case by yawning and closing his eyes during our arguments and, conversely, by leaning forward with interest and smiling and nodding his head during the state's presentations. He set up the verdict sheet sent to the jurors so that a finding for the defendant necessitated only a single quick "no" response, while a finding for the plaintiffs would require the jurors to answer each of a long series of questions a specific way. He then sent the jury out to deliberate at 5 PM on a Friday, with the highly disingenuous words, "Stay as late as you want, and if you don't finish, come back Monday." The judge thereby effectively ensured the jury would take the quick way out by finding in favor of the defendants.

The judge refused us the usual rebuttal time during closing arguments. He excluded the other Spanish-speaking parents of boys also circumcised at Elmhurst Hospital, who would show a pattern and practice of discrimination by the hospital. Judge Friedman thereby effectively made it impossible for us to prove our federal claims, which is why we were in federal court in the first place.

The judge barred Doctors Cold and Van Howe from testifying. In Cold's case, it was because he does not perform circumcisions, and has no training in ethics or law. Bob Van Howe actually showed up at the courthouse, but still was not allowed to testify. Judge Friedman accepted the defense's contention that his statistical evidence relating to circumcision would be prejudicial (even though he has a Masters degree in statistics, along with medical expertise), yet refused to delay trial due to extremely prejudicial testimony of pregnant doctor. He also refused to sign an order denying us a stay, thereby effectively barring us from appealing his decision. Cases hang on such fine points. Van Howe was not even allowed to testify to impeach Dr. Bernstein's evasive, lying testimony.

The judge in the Anastasian case ran a tight ship. She was fair, but not willing to make exceptions.

Juries

Juries are an interesting relic of our legal culture's evolution from its British roots. In both the Pappas and Anastasian cases, juries did not give our clients what we felt they were due. There are understandable reasons why juries withhold their compassion.

In the Pappas case, visiting Judge Friedman hamstrung our ability to screen the jurors by barring us from directly interviewing the jurors. All questions for the jurors had to be passed through him and his magistrate despite the fact that other visiting judges allow direct juror questioning. Talking to the jurors directly allows development of a rapport, a relationship. This inevitably assists plaintiffs in educating a jury regarding their loss.

Judge Friedman evidenced a breathtaking willingness to prejudice the jury against us. He prevented the jury from even seeing critical evidence, excluded crucial expert witnesses from testifying, and pressured the jury by setting up the verdict sheet and timing events so that jurors were in a position where they could only avoid missing more days of work by quickly deciding against the Pappas boy. Perhaps most damagingly of all, he vividly telegraphed to the jurors how they should decide the case. In a virtually inevitable result, the jury decided against the Pappases at the trial court level.

In the Anastasian case, our jury had a lower Armenian representation than we expected. We did win $20,000, obviously a substantial sum of money and a sign the jury appreciated the harm William had endured. Yet this amount was too low to cover costs and expenses. Moreover, the jurors gave nothing to Laura. A juror we spoke with after trial stated that they had trouble with William's lifestyle, presumably meaning his mistress (even though Laura knew and agreed). Also, it may have been hard for them to reconcile the client's arguably conflicting statements of 95% sexual sensitivity loss and that sex was still very good. Since the Anastasians never reimbursed us after trial, as they were required to do by our representation agreement, my co-counsel and I ended up out several thousand dollars each.

Courts and Procedures

We made an early decision to try the Pappas case in federal court and start to build a federal record regarding genital integrity. Ultimately equal protection of the right to genital integrity for males and females will have to be adjudicated by a federal court. Federal courts tend to be advantageous to plaintiffs, with more sophisticated judges and juries, less subject to local pressures and prejudices. Naturally, we knew we would not be staying in federal court with just a medical malpractice claim. The Fourteenth Amendment of the Constitution has been held in past cases to protect parents' right to make decisions about the care, custody, and control of their children. We, therefore, sued under the Fourteenth Amendment. This approach enabled us to deftly sidestep one common roadblock in circumcision cases — peoples' prejudices and vested interests regarding circumcision. This case was based on the Constitution of the United States of America. On the other hand, a unanimous jury is required in federal court.

We were elated when we survived the defendants' Motion to Dismiss with the late Judge Nickerson. Judge Friedman forced us to brief the issue of whether there was federal jurisdiction, then reaffirmed the previous order regarding federal jurisdiction, "based on the court's own research." Mysteriously, the judge never specified what constitutional issues formed the precise basis for jurisdiction. Friedman may have been trying to help us to reach a settlement, or may have been planning his upcoming hatchet job in advance.

A second major procedural issue was the possibility we seriously explored of transforming the case into a class action. We managed to obtain information on total circumcisions at Elmhurst Hospital over a 13-year period, including 1997, the year of Dennis' circumcision. We interviewed fourteen other potential plaintiffs in detail, usually talking to them in Spanish. We compiled evidence that Elmhurst was targeting Latino boys for circumcision based on their parents' surnames, and also that the circumcisions were being performed without informed "consent." Elmhurst showed a higher incidence of circumcisions of Latino boys than would be expected. For example, in 1997, 26% of Latino boys born at Elmhurst were circumcised. This compares with a baseline circumcision rate for Latino boys in the US of no more than 5–15%.

As mentioned above, Judge Friedman did not allow our other witnesses to testify, holding that their participation would require "mini-trials" to be held regarding their cases. We argued that no mini-trials would be needed, as the issues are streamlined and essentially the same in the different cases, basically with only dates and names varying. We found some authority from case law supporting our interest in bringing in other similar cases. But the judge maintained his position, ironically implying that, if we had fewer witnesses, he might have considered it. And yet, we lacked sufficient witnesses to bring the matters all together in one class action. We were whipsawed between having too few cases and too many. In a sane world, given the concerns and harms common to all, a class action would have had a chance to proceed in some format, in some venue. One usual problem with class actions is the plaintiffs do not have so-called typicality of damages. If a plane crashes, some will die, some will suffer head injuries, and others will suffer internal injuries. Even in an asbestos lawsuit, the level of illness tends to vary. But here we did have typicality, as all potential plaintiffs have essentially the same harm — a lost foreskin.

We wanted to appeal the Pappas case but the clients stopped paying for expenses and then later stopped communicating civilly with us. We still considered footing the considerable costs of appealing ourselves but eventually were regretfully forced to decide against this highly risky and uncertain course.

Even to get the Anastasian case started, several barriers had to be overcome. First, to avoid the one-year statute of limitations, we needed a copy of a letter from Anastasian's prior lawyer to the defendant, advising him of the lawsuit. Obtaining this letter was not as easy as it might have been, though eventually we did get it. Secondly, at that point in time, I was living with my wife in the US territory of Guam, many thousands of miles from California. I had to find a local co-counsel, get him on board with the case, and orchestrate a meeting between him and the Anastasians.

One interesting discussion involved that state's so-called "998 offers," referring to offers to settle a case that can tie the other side's hands. Pursuant to Section 998 of the California Civil Procedure Code, if one side rejects a 998 offer, and the ultimate court award is better for the offering side than the amount of the 998 offer, then the losing side has to pay most of the winning side's expenses and court costs incurred from the offer date on, plus interest. Strategically, therefore, one tries to choose an amount for a 998 offer that is high enough that the plaintiff can live with it if the other side accepts the offer, but still low enough that we should be able to exceed it at trial, thereby making us eligible for the often very significant expenses and costs under Section 998. Our offer to the defendants was in the low six figures, a region we could live with if they paid it and yet that we believed (incorrectly, as it turned out) we could probably exceed at trial. We hoped for punitive damages due to the battery and knew that according to a well-known California Supreme Court case that was still good law,[30] California's $250,000 cap on medical malpractice cases,[31] therefore, was inapplicable. This was medical negligence, not medical malpractice. Of course, the defendants refused our offer. Ultimately William won a somewhat pyrrhic victory at trial, receiving $20,000, though due to the high cost of legal actions, this failed to cover expenses and costs. Laura's claim was denied outright.

Doctors

The Armenian doctor who had circumcised William Anastasian, Dr. Dostourian, was a piece of work himself, having faced over two dozen malpractice lawsuits against him. At first, he blamed the nurse for his own mistake, then eventually recanted. Circumcised while in the Army, he clearly lacked compassion for Anastasian. As a distinguished man of advanced years who spoke softly and was a physician, he probably also obtained sympathy from the jury, despite his history of harming patients and, in this case, blaming subordinates for his own errors.

We briefly contacted and then had to extricate ourselves from an overeager Armenian doctor who was a potential medical expert. He clearly was not too knowledgeable and would not have played well with the jury. We also had to refrain from hiring another medical expert who had liabilities we thought would not play well with a jury, including a criminal history.

One physician, entitled only to a nominal ordinary witness fee, attempted to extract thousands of dollars from us as an "expert witness," which of course he would not be as a percipient witness.

In the Pappas case, Dr. Bernstein was allowed to testify while over eight months pregnant. She demonstrated a remarkable ability not to give us any information at all, while not directly refusing to answer our questions. She managed, at the same time, to manipulate her condition by playing the role of the nice, ostensibly cooperative professional who was also an expectant mother. This performance evidently swayed jurors toward sympathy for her and, by extension, the defendant hospital and city. We objected to the obvious prejudice created by having her testifying in her condition, but Judge Friedman refused us a continuance.

Other Groups — Media, Activists

As an activist who happens to be a lawyer, I am inevitably wearing multiple hats in any such case, and will be working with other activists and often with media, though my primary duty, of course, is to my clients. In the Pappas case, with authorization from the clients, I worked with activists from the National Organization of Circumcision Information Resource Centers (NOCIRC) to prepare a press release.[32] In this case, we also worked closely with a very sympathetic member of the Spanish-language media, whose newspaper featured daily stories about the trial, sometimes on the front page. Television coverage also took place, and I was interviewed in Spanish.

William Anastasian was ostensibly also a strong activist, though, for various reasons, we did not work directly with the media or other activists on his case.

Conclusion

Each case shares the difficulties faced by all litigation, and each also has its own particular pitfalls and strong points. The outcomes of lawsuits hang on the details: a very pregnant circumcising doctor getting the jury's compassion, expert doctors not allowed on the witness stand, pattern and practice witnesses excluded, a judge telegraphing to a jury how to decide a case. Unfavorable jury perceptions no doubt affected the result in the Anastasian case.

General difficulties in bringing circumcision lawsuits include (1) financial risks; (2) procedural difficulties; (3) misconceptions and compassion misallocation among judges, lawyers, jury members, the media, and the general public; (4) constraints unique to circumcision lawsuits that are imposed by statutes of limitation and statutes of repose; (5) need for parental participation in lawsuits; (6) problem of damages not being atrocious enough to justify litigation; and (7) the scarcity of helpful case law. Players whose roles we have explored include clients, lawyers, judges, juries, courts and procedures, doctors, media, and fellow activists. All are important. Cases are always decided holistically. Yet, each is the sum of the parts.

References

1. Gilbert v. Sacred Heart Medical Center, 127 Wash. 2d 370, 900 P.2d 552 (1995).
2. Circumcision Information and Resource Pages. United States Circumcision Incidence [cited 19 August 2006]. URL: http://www.cirp.org/library/statistics/USA/
3. Centers for Disease Control, National Center for Health Statistics. National Hospital Discharge Survey. Series 13, Number 162. October 2006 [cited 20 February 2008]. URL: http://www.cdc.gov/nchs/about/major/hdasd/nhds.htm.
4. Zeldis A. Parental roles in circumcision lawsuits. Attorneys for the Rights of the Child Newsletter, volume 4, no. 1 (Winter 2004), pp. 7–8.
5. Eastern District of New York Case No. CV 98-5459, 1999 U.S. Dist. LEXIS 21308 (E.D.N.Y. 1999).
6. London v. Glassner, California Court of Appeal, 1st District, No. A032040 (unpublished, petition for review denied); *see also* Morris R. The first circumcision case. Truth Seeker 47 (July/August, 1989).
7. Fishbeck v. North Dakota, 115 F.3d 580,581 (8th Cir. 1997).
8. Liptak A. Circumcision opponents use the legal system and legislatures. New York Times; 23 January 2003 [cited 16 August 2006]. URL: http://www.cirp.org/news/newyorktimes 01-23-03/
9. Eastern District of New York Case No. CV 98-5459, 1999 U.S. Dist. LEXIS 21308 (E.D.N.Y. 1999).
10. Noe v. Kaiser Foundation Hospitals. 435 P2d 306 (Ore 1967).
11. Richard v. Louisiana. 526 So. 2d 1237 (La. 1988).

12. Felice v. Valleylab, Inc. 520 So. 2d 920, *cert. den. (La.),* 522 So. 2d 562, *cert. den. (La.),* 522 So. 2d 563.
13. Doe v. Shrom. 664 A.2d 102 (Penn. 1995).
14. Wilson v. Lockwood. 711 S.W.2d 545 (Mo. 1986). A directed verdict was upheld for a second defendant under a lack-of-consent claim where parents conceded receiving brochure explaining risks despite lack of verbal explanation from hospital.
15. Johnson v. Hammond. Ohio App Lexis 1226 (1988).
16. McNabb v. Mason, 148 Ind App 233, 264 NE2d 623 (Ind. 1970).
17. Sweet v. Sisters of Providence in Washington, 893 P.2d 1252 (Alaska 1995).
18. Valentine v. Kaiser Foundation Hospitals. 15 Cal. Rptr. 26 (1961).
19. In the Matter of the Marriage of Boldt and Boldt, Oregon Supreme Court Case No. S054714 (January 25, 2008) [cited 20 February 2008]. URL: http://www.publications.ojd.state.or.us/S054714.htm.
20. Beatty v. Charkravorty. 1997 US Dist. Lexis 14925 (1997).
21. Johnson v. Colp. 300 NW 791 (Minn. 1941).
22. Bates v. Newman. 264 P2d 197 (Cal 1953).
23. Walker v. Skiwski. 529 So.2d 184 (Mississippi 1988).
24. Kalina v. General Hospital of Syracuse. 220 NYS2d 733 (1961), *aff'd,* 235 NYS 2d 8089 (1962).
25. Lupo V. Family gets $2.75 million in wrongful surgery suit. Lake Charles American Press, 28 May 1986 [cited 18 August 2006]. URL: http://www.cirp.org/news/1986.05.28_lawsuit/.
26. Waddell D. Sex operation pilot settles for £800,000. (London) Telegraph, 27 November 1998.
27. Associated Press. $1.4 Million awarded for botched circumcision. Santa Cruz Sentinel, 9 June 2001 [cited 18 August 2006]. URL: http://www.cirp.org/news/santacruzsentinel06-09-01/
28. Reinert S. $80,000 settlement to circumcised boy. Patriot Ledger, 5 January 2001 [cited 18 August 2006]. URL: http://ledger.southofboston.com/articles/2001/01/05/news/export8626.txt.
29. National Organization of Circumcision Information Resource Centers and Attorneys for the Rights of the Child. Doctors and Hospital Settle Circumcision Lawsuit. Men's News Daily, 29 April 2003 [cited 22 August 2006]. URL: www.cirp.org/news/mdnewswire04-29-03/
30. Cobbs v. Grant, 8 Cal. 3d 229 (California Supreme Court 1972).
31. Medical Injury Compensation Reform Act of 1975 (MICRA), California Civil Code Section 3333.2
32. National Organization of Circumcision Information Resource Centers. $25 million civil rights circumcision case to protect Spanish speaking families starts trial Monday in federal court. 31 July 2003 [cited 16 August 2006].URL: http://www.arclaw.org/news/0000001a.htm.

Winning and Losing on the Circuit

David J. Llewellyn

Abstract Eleven years of litigating wrongful circumcision and other genital injury cases has taught me a great deal about the strong emotions individuals have in regard to their own genital state or the genital state they believe is "correct." These emotions are often so overwhelming that they overcome any sense of justice or fairness when it comes to the genital state of others. Estranged parents often try to circumcise a child at a later age to satisfy either their sense of the way a penis should look or to "get back" at their ex-spouses. A sense of power and entitlement is often involved. Upon occasion, insignificant, transitory inflammation of the penis can lead to a demand for circumcision. Sometimes it seems a new spouse's ideas provide the impetus for seeking the circumcision of a boy in the middle of his youth. Courts often are not well prepared to deal with these issues, particularly since many doctors are still ignorant of the very effective medical modalities available to treat foreskin problems without surgery and often recommend circumcision when it is really contraindicated. The recent Chicago case involving a demand for circumcision of an 8-year-old boy is reviewed and considered.

As a trial lawyer from the American Deep South, I have practiced in the area of civil litigation for almost 27 years. Since 1995, I have handled a variety of circumcision-related cases throughout the United States. Eleven years of litigating wrongful circumcision and other genital injury cases has taught me a great deal about the strong emotions individuals, whether they be jurors or judges, often have in regard to their own genital state or the genital state they believe is "correct." These emotions are often so overwhelming that they overcome any sense of justice or fairness when it comes to the genital state of others. Usually the attitudes of the "triers of fact" are not known and may never be known. However, the possibility that those feelings strongly favor circumcision must be factored into a lawyer's analysis of a genital injury case and taken into account when planning strategy for presentation of evidence and argument or the case may well be lost. Careful planning does not always ensure success even by the most skilled counsel. However, an appreciation of the emotional nature of circumcision is paramount. When these emotional factors are

---

A trial lawyer with the firm of Johnson & Ward in Atlanta, Georgia, USA

G. Denniston et al. (eds.), *Circumcision and Human Rights*,

taken into account in advance and adequately countered by careful presentation and argument, it is quite possible for a plaintiff to prevail in a circumcision case.

It is, I believe, generally accepted that male neonatal circumcision is a social norm in the United States. As I discussed at length at the Eighth International Symposium on Circumcision, Genital Integrity, and Human Rights in Padua, Italy,[1] it is unlikely that the legal system in this country will begin to treat male genital injury cases with complete fairness until we have gone well beyond the "tipping point" and arrived at a situation where male genital integrity is the social norm. We are not yet there in most of the country.

Recently, I ate lunch at an oriental restaurant. Upon cracking open my postprandial cookie, I discovered my fortune. It read: "People willingly believe what they wish." Nothing could be truer. Thus, the well-known pundit David Brooks recently observed:

> Walter Lippmann got to the crux of the matter 65 years ago. People don't become happy by satisfying their desires, he said. They become happy by living within a belief system that restrains and gives coherence to their desires:
>
> Above all the other necessities of human nature, above the satisfaction of any other need, above hunger, love, pleasure, fame — even life itself — what a man needs is the conviction that he is contained within the discipline of an ordered existence.[2]

In the United States, our ordered existence imposes the discipline of neonatal male circumcision. It is this discipline, coupled with an ingrained sense of conformity and general insecurity about sexual matters, that makes it so hard to use the legal system to effect any real meaningful change in how we treat wrongful circumcision cases and other male genital injury cases.

As one commentator recently put it on the Circumcision Debate board at motheringdotcommune, the underlying psychological issue that perpetuates male circumcision is male insecurity. "'All Penis's [sic] must look like mine or I am a freak[.]'…'It was done to me, are you saying my penis is inferior??? Impossible.'"[3]

Josephine Marcotty, in an article entitled "Why circumcise?" in the August 16, 2006, issue of the Minneapolis — St. Paul, Minnesota *Star Tribune,* observed:

> The choice is largely driven by cultural preference. In the United States, most boy babies are circumcised because most American men are. That, and the fact that circumcision is usually paid for by health insurance, is why the practice continues, said Eli Coleman, a professor and director of the Program on Human Sexuality at the University of Minnesota.
>
> 'We are all trying to look like a man … is supposed to,' he said ….
>
> 'It's not what the girls think, it's what the boys will think,' said Coleman. 'It's that father and son will go into the shower and be comparing their penises, that this is what a good man looks like.'[4]

This is not so different from Romeo B. Lee's observations about Filipino circumcision. In the Philippines, "men submit to the procedure (circumcision) because they want to be understood as 'men' and 'circumcised' (and not consequently as 'different,' 'cowards,' or 'homosexual')."[5] "Filipino male circumcision therefore appears to be a social phenomenon, propelled by the individuals' need to conform to a centuries-old tradition and to acquire, through that tradition, a range of masculine-related traits, capacities, and opportunities. Fulfillment of this need is perceived as subsequently leading to the attainment of psycho-social health — a sense of personal wellbeing, recognizing that one has adhered to a community-wide practice; one has become a 'man' and 'masculine'; and more importantly, one has been socially accepted."[6] "The structural and social embedded-ness of the male circumcision represents a major barrier to change."[7] So, also, is it in the United States.

Vincent Bach in his trenchant Internet essay "The Vulnerability of Men"[8] says:

> First of all, you need to understand that circumcised men are cornered on this issue. They were circumcised without their consent and have no inherent knowledge of what being intact is like. Even though they rarely will discuss the issue, they are keenly aware that they have been surgically altered in a very private way. There are several ways for a man to deal with this issue but the safest way, psychologically speaking, is to believe at all cost that the surgery performed on them was an enhancement and is preferred by women. Confirmation of this belief is essential to their sexual self-image....
>
> I think it's important to acknowledge that it's perfectly understandable that our circumcised friends react this way. Men who have been circumcised have an extremely difficult dilemma. For them to acknowledge that the practice is unnecessary and harmful means that they must acknowledge a painful personal reality. For that reason, circumcised men can be forgiven if they don't want to lead the parade in the fight against routine infant circumcision. I can empathize and, therefore, understand completely why so many men will voluntarily offer their sons up for the same procedure without giving it a second thought. To do otherwise opens them up to some vulnerable feelings that can be most unpleasant.

These feelings are often expressed vehemently whenever circumcision is challenged. For example, the men's magazine, *Playboy,* in "The Playboy Advisor" section of its May 2006 issue, when asked what men think about being circumcised said:

> Most men had no say in the matter. We see no reason for the procedure, nor does the American Academy of Pediatrics, which doesn't recommend it. Some research suggests that circumcised infants have fewer urinary tract infections and that circumcision may help prevent HIV transmission. But these risks can be addressed by less radical means, such as regular washing and using condoms. The more we learn about the complexity and function of the prepuce, the more a tragedy it seems to lop it off, even as a religious ritual. It has long been dismissed as a useless piece of skin, but on closer examination it appears to be similar to the tissue between the facial skin and the mucosa inside the mouth. For that reason, notes David Gallaher, who has written a history of the surgery, the nerve endings of the foreskin have been compared to those in the fingertips and lips. To cut either of those parts from an infant would be considered barbaric.[9]

The replies to this sound advice were printed in the September 2006 issue. Shouted D.M. of Benton, Kansas:

> Your anticircumcision diatribe in May is silly and wrong. Despite what David Gallaher claims in his book, there are no highly sensitive nerve endings in the foreskin; they are in the glans, just as a woman has nerve endings in the clitoris rather than the labia minora. The foreskin is merely an extension of the skin covering the shaft and has virtually no feeling. The American Academy of Pediatrics, which in 1999 decided not to recommend circumcision, caved to political pressure from a rather bizarre group of people claiming the procedure is on a par with removing the clitoris and/or labia. The data are clear that women partnered with circumcised men have fewer vaginal infections and lower rates of cervical

> cancer. The foreskin probably had a protective function at some point, just as the hymen probably served as a barrier to fecal contamination. Indeed in third world countries where hygiene is a luxury, they may still serve these functions. Otherwise the data suggest circumcision is best. Incidentally, I teach a college course in human sexuality, and when this issue comes up I have yet to hear a female student say she finds an uncircumcised penis more attractive.[10]

In a similar vein, a mother wrote: "But wouldn't the pain of rejection at the hand or mouth of a girl unsure of what to do with or grossed out by an unfamiliar foreskin sting far worse (than the pain of circumcision)?"[11]

These reactions point out that the desire to conform to the social norm of male circumcision is alive and well in the U.S. But they also point out that the tide is beginning to shift, albeit slowly. As a sage once pointed out, whenever social change is attempted, it is at first subjected to ridicule and then to anger. We are now clearly at the stage where opposition to circumcision is growing and is so threatening to the social norm that it is opposed with anger rather than ridicule, which was formerly employed.

The anger of circumcision proponents extends even to the baby advice literature. Recently, a book, entitled *Baby 411*, has been widely praised and touted on national television. It is filled with exaggerations in regard to neonatal circumcision. For example, it states: "The practice of circumcision…became common in the United States in the late 1800's for hygienic reasons. It has continued to be the 'mainstream' choice for most American boys. …"[12] Advantages discussed are hygiene (less HPV), UTI's, penile cancer, and cervical cancer. In regard to HIV, it says "Uncircumcised men are more likely to have the HIV virus and infect their partners. Why? Because the area under the foreskin makes a nice spot for the virus to set up housekeeping."[13] In regard to sexual pleasure, the book states: "It is hard to assess how circumcision affects sexual pleasure. I don't think either group of men is complaining. While there are some groups out there who care deeply about what you choose to do with your son's foreskin, I think it is a very personal decision without a clear answer besides personal preference."[14] It goes on to warn, apparently in regard to authoritative anti-circumcision websites such as www.cirp.org, "Note to Internet users: Google 'circumcision' at your own risk. You will get bizarre results from fringe groups with their own agenda."[15] Then it summarizes with this: "Bottom Line: The whole procedure takes about 60 seconds. Yes, it hurts. But most babies go to sleep for a few hours and wake up happy to see you. For more information on circumcision, check out these web sites: medicirc.org (Dr. Edgar Schoen's website[1]); circinfo.net (Brian Morris' website, which states 'The fact that

[1] Dr. Edgar Schoen is a pro-circumcision advocate who contends in regard to the adoption of circumcision in the United States, "It's as American as baseball, the stars and stripes, apple pie — and circumcision." The title of the book in which this quote is found is Ed Schoen, MD on Circumcision: Timely Information for Parents and Professionals from America's #1 Expert on Circumcision. In it he attacks the contention that 85% of the world does not circumcise males by stating, "I don't think you are helping our cause with this statement, folks. The American public doesn't take well to the idea of using the rest of the world as a model. U.S. parents don't want to follow the health practices of millions of 'intact' Hindus in the ghettos of India, or of the hordes of uncircumcised peasants in China. We feel that the U.S. is the medical leader of the world, and rightly so. The rest of the world usually follows us, not the other way around." See Schoen, E. Ed Schoen, MD on Circumcision: Timely Information for Parents and Professionals from America's #1 Expert on Circumcision. Berkeley, CA: RDR Books. 2005: 52, 123.

it [circumcision] is still popular must mean there is something in it! Interestingly, in some places, such as Madagascar, circumcision is 100% regardless of religion, and the reason is actually dictated by the women, who maintain that circumcised sex is 'longer, stronger and cleaner.' "[16]), and circumcision.cjb.net (now www.aboutcirc.com) which screams: The basic choice is this: modify nature for better health, or leave it alone?

This insecurity about the circumcised penis often leads to suits, such as one the firm of Lake, Toback & D'Arco of Chicago, Illinois, and I have recently undertaken, in which divorced parents are in court disputing over the attempted circumcision of a healthy 8-year-old boy for balanitis.[17] [See Appendix I] Although the parents agreed not to circumcise their son at birth, they are now divorced. The mother wants to circumcise the boy because of a few bouts of balanitis; the father opposes the circumcision as unnecessary and harmful. We took great care in crafting our written closing argument that sets forth the father's view of the facts and applicable legal principles. Since we did not know the personal feelings of the court in regard to circumcision and since we were well aware of the prevalence of the emotional factors I have just discussed, we stuck to what we knew we had proven: that circumcision was unnecessary in this case at the present time, that circumcision is irreversible, that circumcision is potentially psychologically damaging to an older boy, that circumcision at this age has inherent risks, and that the alleged balanitis that was the basis upon which circumcision was sought is almost always curable by the application of betamethasone cream, which has never been prescribed for the child. We tried to show that the child did not need a circumcision and thus emphasize the "needs of the child" rather than merely relying upon the vague term "best interests of the child," even though we used that obligatory term. We had proven through witnesses the nature of the foreskin and its value so as, we hoped, to demonstrate to the court's satisfaction that a circumcision would be damaging to the boy. Yet, the emphasis remained upon the fact that the child did not require (i.e., "need") a circumcision for present good health. This argument should enable the court to decide the case in favor of non-circumcision without finding that circumcision is harmful in and of itself. Incidentally, while the court-appointed child's guardian contended that our witnesses were too biased against circumcision, he did agree that at the present time circumcision was not needed by the child. Therefore, he recommended that the child be left intact. A decision on this case is not expected until after this symposium.[2]

[2] The case was decided on October 24, 2006. In his Order of that date Judge Jordan Kaplan concluded "that the evidence was conflicting and inconclusive as to any past infections or irritations that may have been suffered by the child." He found "that the medical evidence as provided by the testimony of the expert witnesses for each of the parties is inconclusive as to the medical benefits or nonbenefits of circumcision as it relates to the nine year-old child of the parties." [The boy had turned nine during the pendency of the case.] Judge Kaplan also noted, "the injury to the Child as a result of an unnecessary circumcision would be irreversible." After indicating that he had weighed the "issues of possible psychological and physical harm to the Child," he stated "Circumcision is an extraordinary medical procedure as it relates to a nine year-old child." The court, therefore, issued an injunction prohibiting the child's circumcision before he reaches the age of majority when he can make his own decision, absent the development of a substantial change in circumstances.

While wrongful circumcision cases, i.e., those cases where circumcision is performed without the consent of the parents of a minor or the consent of the patient himself if an adult, are different in nature from injunctive actions involving alleged medical necessity, the successful prosecution of them also requires attention to the emotional factors I have discussed. Usually, in such cases, I endeavor to overwhelm the defense with the indisputable evidence of the loss of the protective function of the foreskin, the loss of the ridged band of the foreskin with the consequent modification of the sexual experience in masturbation as well as vaginal intercourse, the pain of the circumcision itself, the pain experienced during the healing period, the lowering of the child's pain threshold if the plaintiff is a baby, and the like. Despite the difficulties often encountered in convincing the defense that these facts are true, I have been quite successful in obtaining satisfactory results for my clients. I recently reached a settlement of $100,000 for an infant wrongfully circumcised in Virginia and a settlement of over $125,000 for a Kentucky boy who was needlessly and wrongfully circumcised without parental permission while undergoing a cystoscopy at the age of five. Slowly but surely the value of the foreskin is being recognized and its loss is being properly compensated.

In conclusion, it is my belief that circumcision cases can be successfully concluded so long as the lawyer takes into account the potential prejudices in favor of circumcision, presents sufficient evidence of the scientific and indisputable facts regarding the nature and function of the foreskin, the dangers of circumcision, its harms, and its lack of medical necessity, and then sticks to those facts in the arguments made to the defense, the jury, and the court.

References

1. Llewellyn, D. Strategies for litigation, Chapter 19 in Denniston, G., Gallo, P., Hodges, F., Milos, M., Viviani, F. Bodily Integrity and the Politics of Circumcision: Culture, Controversy, and Change. New York. NY: Springer Science + Business Media, Inc; 2006.
2. Brooks, D. Any cultural change will be slow process. The Atlanta Journal Constitution, August 15, 2006, p. A7.
3. ShaggyDaddy, The Case Against Circumcision Discussion Board.. www.mothering.com/discussions/showthread.php?t=498823, accessed August 10, 2006.
4. Marcotty, J. Why Circumcise? Star-Tribune, Minneapolis-St. Paul, MN. www.StarTribune.com/circumcision081606, accessed August 10, 2006.
5. Lee, R. Filipino experience of ritual male circumcision: Knowledge and insights for anti-circumcision advocacy. Culture, Health & Sexuality, May–June 2006; 8(3):225–234.
6. Ibid.
7. Ibid.
8. www.stopcirc.com/vincent/vulnerability _of_men.html, accessed July 19, 2006.
9. The Playboy Advisor. Playboy, May 2006; 53(5): 35–36.
10. D.M. in The Playboy Advisor, Playboy, September 2006; 53(7): 39–40, 39.
11. N.M., Ibid.
12. Brown, A. and Fields, D. Baby 411: Clear answers & smart advice for your baby's first year, 2nd edition. Boulder, CO: Windsor Peak Press; 2006:14.

13. Ibid.
14. Ibid, 15.
15. Ibid.
16. Accessed August 25, 2006.
17. In Re the Former Marriage of: Adriana Schmidt, Petitioner, and Marian Niznik, Respondent, No. 00 D 18272 cons. 00 D3 31904, Circuit Court of Cook County, Illinois, County Department — Domestic Relations Division.

The MGM Bill
A Legislative Strategy for Protecting US Boys from Circumcision

Matthew Hess

Abstract The goal of MGMbill.org is to pass federal and state laws that would protect boys in the USA from circumcision the same way that girls are protected under existing female genital mutilation statutes. Our MGM Bill proposals have been submitted to each member of Congress and the California State Legislature three times, and in 2006 the effort was expanded to include 14 additional state legislatures.

In addition to encouraging the American public to write lawmakers in support of our proposed legislation, MGMbill.org is lobbying Amnesty International and the United Nations to classify male circumcision as a human rights violation

Federal Male Genital Mutilation (MGM) Bill Proposal

The MGM Bill effort is an attempt to protect boys in the United States from medically unnecessary circumcision by enacting legislation. Our federal MGM Bill proposal is written as an amendment to the current US Female Genital Mutilation law (§18USC116), and has been submitted to every member of Congress three times (in 2004, 2005, and 2006). No one in Congress has agreed to sponsor the bill. However, some lawmakers did respond in writing. Here are excerpts from three of those replies:

Rep. Virgil Goode (R-Virginia, 5th District)

> I want to see genital mutilation and other forms of mutilation outlawed to the fullest extent possible and will gladly show every consideration to the Genital Mutilation Act of 2005 and other legislation that deals with this topic.

Senator Orrin Hatch (R-Utah, Health Committee Member)

> No definitive information has come forth as to the long-term emotional effects that neonatal circumcision may have.... Ultimately, I believe that parents should decide what is best for their children.

Founder and Director MGMbill.org, San Diego, California, USA

G. Denniston et al. (eds.), *Circumcision and Human Rights*,

US Rep. Sherrod Brown (D-Ohio, 13th District, Subcommittee on Health)

> Should the medical community determine male circumcision is unsafe and recommend the procedure not be performed, I would support their recommendation.

State MGM Bill Proposals

In 2004 and 2005, a state-level MGM Bill proposal was submitted to all 120 members of the California State Legislature. In 2006, that effort was expanded to include an additional 14 state legislatures through a volunteer state office program, reaching 2,195 state lawmakers in all. One early response from California State Senator Christine Kehoe (D-39th District) advised that "...you will need to obtain the support of numerous health and human rights organizations that are widely recognized and respected," which led to the formation of an endorsement campaign that now has nine signers.

Massachusetts MGM Bill

The state MGM Bill submissions resulted in one formal sponsorship (in 2006) by Massachusetts State Senator Michael W. Morrissey (D-Norfolk), on behalf of his constituent Charles A. Antonelli. Senator Morrissey initially requested that a religious exemption be added, but we countered that the federal Female Genital Mutilation statute had no such exemption, and that including one for males would be gender discrimination. The Senator then agreed to leave it out, and the bill was filed on April 13th, 2006, as Senate Docket #2621.

Unfortunately, the docket was filed after the normal legislative deadline, meaning that both the Senate and House Rules Committees must agree by a two-thirds vote to release the bill to the Joint Committee on the Judiciary for a hearing. At MGMbill.org's request, Senator Morrissey's office successfully petitioned the Senate Rules Committee to release SD #2621. The docket is now in the hands of the House Rules Committee, where we are attempting to get it released via a petition that will be passed around this room.

Amnesty International

MGMbill.org[1] is encouraging mainstream human rights organizations to recognize forced circumcision as a human rights violation. Last year, we submitted two MGM resolutions at the 2005 Amnesty International USA Western Regional Conference in San Francisco, California. Those resolutions were both voted down, but we will make another attempt in 2006 at the next Western Regional Conference in Tucson, Arizona. MGMbill.org also exhibited at Amnesty International's 2006 Annual General Meeting in Portland, Oregon.

American Academy of Pediatrics (AAP)

Because the American Academy of Pediatrics (AAP) holds so much influence over US circumcision policy, MGMbill.org is lobbying 11 top AAP officials through an email campaign, urging them to adopt our proposed MGM Policy Statement. Overall, the relationship is cordial, and our messages are reaching the very top levels of this organization.

United Nations

With the current frenzy surrounding male circumcision and AIDS, MGMbill.org is directing a similar email campaign toward three United Nations agencies: the World Health Organization (WHO), the United Nations Children's Fund (UNICEF), and the United Nations Population Fund (UNFPA). The latest WHO statement on male circumcision emphasized "voluntary" circumcision, so it's possible that our correspondence is having an impact.

Press Releases and Events

To reach out to the public and the media, MGMbill.org has issued fourteen press releases through PR Web and other news bureaus since December 3, 2003. We also sponsor occasional events, the latest of which was an MGM contingent in the 36th Annual LGBT Pride Parade in San Francisco, California.

Printed Materials

To help support our campaign, MGMbill.org publishes a variety of promotional materials, including brochures, postcards, and other handouts.

Update (June 28, 2007) — The MGM Bill proposals were resubmitted to Congress and 16 state legislatures in January 2007. The 2006 Massachusetts MGM Bill docket expired in the House Rules Committee but Senator Morrissey's office reintroduced it before the 2007 filing deadline, and the bill is being reviewed by the Joint Committee on the Judiciary. Also, the MGM resolution presented at the Amnesty International 2007 Western Regional Conference was voted down, and a Massachusetts MGM Bill abstract submitted to the AAP 2007 National Conference & Exhibition was rejected. MGMbill.org's proposed legislation now has 22 endorsements.

Reference

1. MGMbill.org — http://www.mgmbill.org.

Circumcision in European Countries

Review of the Possible Annual Number of Laws and Regulations and of Economic Aspects

Yngve Hofvander

Abstract About 20 million boys are circumcised annually. Of those, about 10 million are Muslim, about 1 million are Anglo-Saxon, mainly from the USA. In the medical literature, the focus is almost entirely on circumcision in the USA, while very little is written and studied from other industrialized countries. In Europe, there has been an influx of many millions of immigrants from the Middle East, Africa, and Southern Asia — all areas that practice circumcision in boys (and, to some extent, in girls).

In Sweden, it is estimated that some 3,000 circumcisions are performed annually on Moslem boys but only some 40 on Jewish boys. For four years, we have had a law regulating circumcision but this can be applied only to Jewish circumcision. For the rest, we know nothing of where, how, by whom, or the cost or complications.

I am attempting to compile whatever information can be obtained from 10 European countries that are known to have a large influx of Muslim immigrants and, thereby, probably tens of thousands of circumcisions annually. Who is performing circumcision, where, how, at what cost, who is paying, the number of complications, attitudes from the original population, etc., are the elements I am trying to determine.

My findings will be presented and discussed, including what can be done to achieve a change of attitude and to limit the present high rate of circumcision and, thereby, unnecessary suffering.

Introduction

My interest in the phenomenon of male circumcision started during my seven years as a pediatrician at the Ethio-Swedish Pediatric Clinic in Addis Ababa, Ethiopia. During those years, I was confronted daily with the results of botched circumcisions and also with a wide range of other traditional "operations," such as cutting of the uvula, extraction of non-erupted canine teeth, burning of eyelashes, outer ears, chest, and abdomen, female circumcision (at one week), etc., resulting in the killing or

Professor, International Child Health, Uppsala University, Sweden

G. Denniston et al. (eds.), *Circumcision and Human Rights*,

mutilating of large numbers of children. This happened some decades ago, but seemingly not much change has taken place.

It, therefore, was not surprising to read in the *South African Press Association* (SAPA), May 1, 2006, the following:

> The circumcision season in the Eastern Cape province in South Africa ended with 22 boys dead: 5 boys had penile amputation, 239 were admitted to hospital for severe infections, and 536 boys were rescued from the bush where they were left to die. Fifteen circumcisers were arrested.

These statistics confirmed that traditional circumcision, or rather, male genital mutilation (MGM) still is "alive and kicking."

The notice in SAPA was published in a fairly remote part of the paper and apparently did not cause much stir. The same was true for our Ethiopian reports or publications: the Ministry was only moderately interested.

However, Europe is presently seeing a large influx of immigrants from the Middle East and Africa, bringing with them customs that may include traditions of the sort cited above and, in particular, male and female circumcision. The latter is outlawed in most countries but is still practiced. MGM, however, is practiced on a large scale with surprisingly little legal reaction or reaction from the medical establishment.

The aim of this study and resulting paper was to investigate the rate of MGM in 10 large European countries (all, except Norwegian members of the EU), the legal status of MGM, by whom it is practiced, to what extent state/community-employed doctors were involved, and also who pays for the operation.

The Jewish and Muslim Populations in 10 European Countries

Basically, MGM is practiced worldwide by Jews (approximately 100,000 annually), by Muslims (approximately 12 million), by Africans in traditional settings (approximately 9 million), and neonatally in certain Anglo-Saxon countries, mainly the USA (approximately 1 million annually). In addition, a large part of the South Korean population today is circumcised as a result of influence from the earlier US military intervention.

Although the USA has only a fraction of the total number of circumcisions, it has a total domination of the published papers in the medical literature. Only a minority of what is published concerns the Muslims of African MGM.

Initially, contact was made with embassies representing the 10 European countries (see Table 1) in order to establish contact with the respective Ministry of Health. This became a formidable task because only in a few locations was male circumcision recognized as a medical entity, diagnosis, problem, or responsibility. It seemed to be "hidden" or, at most, something that was on nobody's desk. This indicated that, if anything, MGM has a long way to go before being recognized as something with which we must deal.

The population in the 10 countries is seen in Table 1. Thus, the population totals about 345 million as compared to about 290 million in the US.

Table 1 Population in 10 selected European countries

Country	Population/millions
Denmark	5,431
Finland	5,249
France	60,496
Germany	82,689
Italy	58,093
Holland	16,299
Spain	42,064
Sweden	9,041
UK	59,668
Norway	4,621
Total	**344,651**
USA	290,000
World	6,463,063

Source: Geolive Home page, 2006.

Table 2 The Jewish population in the 10 countries and the proportion in relation to the total population

Country	Jewish population	%
Denmark	7,000	0.13
Finland	1,100	0.01
France	610,000	1.0
Germany	110,000	0.13
Italy	30,000	0.05
Holland	33,000	0.2
Norway	1,200	0.03
Spain	48,000	0.12
Sweden	18,000	0.2
UK	300,000	0.5
Total	**1,158,300**	
USA	5.9 million	2.0
Israel	5.0 million	80
World	14.6 million	0.23

Source: Jewish Virtual Library, 2005.

The Jewish population in these countries is seen in Table 2. It will be seen that the proportion is a fraction of 1%, with France having the highest rate of a little more than 1 million.

The Jewish population in the 25 remaining European countries is seen in Table 3, and again totals about 1 million.

The Muslim population in the 10 countries is seen in Table 4.

The estimated population is usually given as a range because no reporting to the authorities is permitted or practiced. Thus, the Muslim population is about 13 times greater than the Jewish one and constitutes a larger part of the total population. Most are in France and Holland, with more than 5%.

Table 3 Jewish population in remaining 25 European countries

Austria 8,184	Macedonia 100
Belgium 5,821	Moldavia 31,187
Bosnia 1,006	Poland 24,996
Bulgaria 2,300	Portugal 739
Croatia 1,798	Rumania 6,029
Check 3,072	Serbia/Mon 1,732
Estonia 1,818	Slovakia 3,041
Greece 5,334	Slovenia 100
Hungary 60,041	Switzerland 14,978
Ireland 1,204	Russia 717,101
Latvia 9,092	Ukraina 142,276
Lithuania 3,596	Uzbekistan 17,453
Luxembourg 655	
Total	**1,063,653**

Table 4 Muslim population in 10 European countries and % of population

Country	Muslim population	%
Denmark	17,000–190,000	3
Finland	20,000–26,000	0.5
France	3.6–5.0 million	6–8
Germany	3.3–3.4 million	4
Italy	824,000–988,000	1.5
Holland	750,000–945,000	5.8
Spain	607,000–800,000	1.75
Sweden	250,000–350,000	3.5
UK	1.6 million	2.7
Norway	80,000	1.8
Total	**13,379 million**	

Source: National Focal Point (NFP) Report — EU.

Assuming a fertility rate among the Jewish population of 10/1,000 and 20 in the Muslim (ref. Jewish Virtual Library and EU National Focal Point, NFP Report 2005) the number of boys born annually in these two groups would come to about 5,700 and 122,000, respectively. It should be pointed out, however, that these figures are based on rather crude estimations. It should also be pointed out that, other than the religious groups mentioned, for example, African boys also might be circumcised. The extent of this is unknown.

As a minimum, thus, close to 130,000 boys would constitute the target group for MGM in the 10 countries. As is understood, this is most certainly an underestimation. The Muslim population in Eastern and Southern Europe is large and, in some of the former Soviet republics dominating, as it is in Yugoslavia and adjacent countries.

Laws Regulating Male Circumcision

None of the target countries, except Sweden, has any laws or official regulations concerning MGM. In Sweden, a law was passed by the Parliament in 2001, giving rabbis the right to perform the operation under surgical conditions, under supervision by a doctor or a nurse, provided anesthesia was given, and only on infants under two months. This law focused entirely on Jewish boys who are cut on the eighth day. It was understood that Muslim boys were to be cut by a doctor and not under supervision of authorities (nearly all complications, including deaths, have taken place in this category).

Annually, in Sweden, about 40 Jewish boys are circumcised (which equals 40% of all Jewish boys born), and no less than about 3,000 Muslim boys, about two-thirds of whom are probably circumcised during vacation in their home country. (Fig.1)

Guidelines for circumcision are being issued in a few countries, e.g., England and Denmark. In France and Finland, both parents must give consent. Recently, a mother in Finland was sentenced to four years in jail for having had her son circumcised without the father's consent.

In Norway and Finland, where few boys are circumcised, the issuing of a law presently is being discussed.

The Swedish law is presently under review, and a proposal is being put forward to force all government and county doctors to circumcise on parents' request, "free of charge." The fate of this proposal is still uncertain (September 2007).

Who Pays?

In all countries screened, the parents have to pay. However, the cost, in France and England, for example, may be carried by the NHS if it is called therapeutic treatment for "phimosis," which is said to be rather common, and costing from USD 100 to 1,500 (depending on the type of anesthesia).

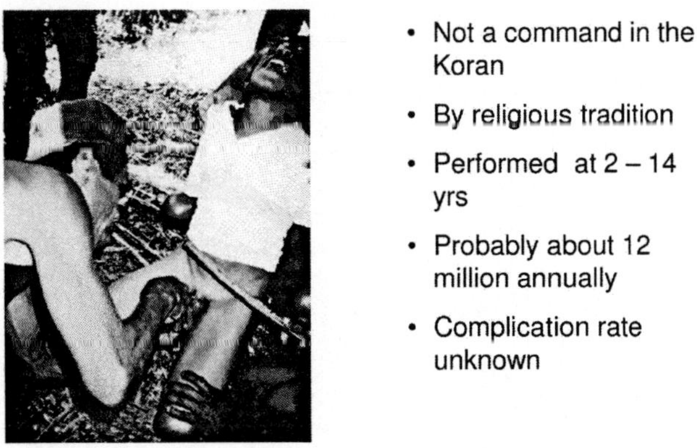

Fig. 1 Muslim circumcision

Where Is It Done?

My contacts in all countries reported there is a massive resistance among hospital doctors to engage in circumcision. Circumcision would have to be done by private practitioners. To what extent circumcision was performed by barbers or traditional circumcisers is virtually unknown. Countrymen from previous-home countries or circumcisers, performing circumcision during temporary vacations, were mentioned as possible "doers." This makes it very difficult to estimate the complication rate because few would be willing to report their contact.

Thus, in all countries there was a more or less total ignorance about where MGM was done, under what circumstances, and at what human cost.

Models

An attempt was made to find out the circumcision status of royalty, focusing on England (ref.:www.circlist.com/rites/british: Do the British circumcise?) Queen Victoria is alleged to have given an order for all royal boys to be circumcised. Prince Charles was circumcised. However, Princess Diana is said to have refused circumcision for her sons, William and Harry.

This has more than gossip interest. If Princess Diana had made public that she did not intend to circumcise her sons, it would have made a tremendous impact on all those who were about to do it.

The famous American pediatrician, Benjamin Spock, let it be known sometime in the midst of his career that he had changed his mind and was against circumcision. Similarly, the following personalities also let it be known that they were intact: Presidents Harry Truman and Ronald Reagan, Martin Luther King, Charlie Chaplin, Clark Gable, James Dean, Yul Brynner, Bing Crosby, Elvis Presley, Frank Sinatra, Ringo Starr, John Lennon, Jack Dempsey, Hugh Hefner (editor, *Playboy*) (reference, personal communication, NOCIRC). To what extent these personalities argued against circumcision is not known.

The United Nations Convention on the Rights of the Child

The United Nations, headquartered in Geneva, has appointed a committee consisting of representatives from 18 member countries. Presently, the chairperson is Dutch. The committee continuously evaluates how the Convention is being implemented in all countries (Implementation for the *Convention on the Rights of the Child*. UNICEF, 1998).

Article 24.3 of the Convention states that "Member states shall take all efficient and appropriate measures in order to abolish traditional customs which may be harmful for the child's health."

Similarly, the *Implementation Handbook*, Article 24, states, "Practices which should be reviewed in the light of the Convention's principles include: all forms of genital mutilation and circumcision."

This last word, so far, has not been on the agenda and, as far as is known, no European country, including Sweden, has raised the question of circumcision (or genital mutilation). It should be noted that quite a few of the countries from which representatives on the Committee come, are practicing both male and female genital mutilation.

Summary and Conclusion

Circumcision in the European context has been and is a sort of "non-issue," in spite of the large number of circumcisions — probably more than 130,000 annually — performed. It is indeed remarkable how little is known about where, how, by whom, and at what cost, in terms of monetary and human suffering, circumcision is conducted.

The present study did not answer all these questions and few officers in the appropriate ministries interviewed had much information to give. And, yet, this is the most common of all "operations," probably causing many more complications than any other operative intervention. But, most complications — in both a short- and long-term perspective — are unknown, and they are left to the boy to endure. Because countrymen and people of the same congregation perform Muslim operations, it would require much effort to report any failure or misconduct.

As for Sweden, which has had a law during the past six years, although this was a failure and is now being revised, it was hoped that it would take a lead and strongly speak out against circumcision. Apparently, this will not happen and — even worse — the government is planning to force all hospital doctors to circumcise babies, free of charge, at the request of parents. It is to be hoped that this will not materialize — that Sweden should not act as a negative model.

But the main problem remains: the majority of the population in European countries is about to give in to an antiquated tradition, which is brutal, harmful, and has no medical indication.

With practically no support or understanding from the UN Committee that implemented the *Convention on the Rights of the Child*, the situation may seem gloomy, with no possibility of an attitude or behavioral change.

And, even worse, the gap will widen between "us and them"; between those who insist on circumcising the genitals of small boys who are unable to exercise their veto and those who abstain in the light of the *Convention on the Rights of the Child*.

"THAT THING"

Portrayal of the Foreskin and Circumcision in Popular Media

Hugh Young

Abstract Popular television has a subtle but significant role in promoting circumcision in the United States. It seems almost obligatory to devote at least part of an episode of every sitcom and soap opera to the topic. The foreskin is commonly denigrated. Contradictory messages are given — for example, that only Jews circumcise babies but all men are circumcised. Pain and harm are minimized or treated as comic. Wherever circumcision is treated as controversial, it is also treated as trivial and inevitable. Talk shows find it good fodder for noisy controversy.

Circumcision occupies a peculiar place in United States culture, being simultaneously ubiquitous, controversial, and a taboo topic of conversation. Thus, to refer to it on television can be simultaneously mundane and daring, a contradiction to which much television programming aspires. It is hardly surprising that references to circumcision maintain a high level of ambiguity: while people may argue about it, the outcome is almost invariably to promote it.

US Television Sitcoms

It is almost inevitable that any US sitcom or soap-opera will sooner or later have an episode or a segment about circumcision. My database[1] now contains 64 shows that say anything coherent about it (Table 1).

Those that have devoted a whole episode to circumcision include *Dharma and Greg*, *Early Edition*, *ER*, *Friends*, *Judging Amy*, *Married with Children*, *The Nanny*, *Off Centre*, *Queer as Folk*, *Seinfeld*, *Sex and the City*, *South Park*, *Thirty-something*, and *Undressed*.

Others have introduced the topic as a subplot throughout an episode, and many more have had a scene about it or mentioned it in passing. (They were gathered by no particular means, so bias in selection is undetermined.)

Independant researcher who created and maintains the human rights website, www.circumstitions.com
Pukerua Bay, New Zealand

G. Denniston et al. (eds.), *Circumcision and Human Rights*,

Table 1 TV sitcoms and soap operas that refer to circumcision

According to Jim	Give My Head Peace (UK)	Off Centre
All in the Family	Grosse Pointe	The Practice
Arrested Development	House	Providence
Angel	Jackass	Queer as Folk (US version)
Bob Paterson	Judging Amy	Scrubs
Caroline in the City	Kids in the Hall	Seinfeld
Cheers	The King of Queens	7th Heaven
Chicago Hope	King of the Hill	Sex and the City
Circumcized [sic] Cinema	Kyle XY	Shortland Street (NZ)
Comedy Inc.	The Kumars at No 42 (UK)	The Simple Life
Crossing Jordan	Ladies' Man	The Simpsons x4
Cybil?	Law and Order: SVU	Six Feet Under
Dawson's Creek	Living in Captivity	South Park
Dharma and Greg	Married with Children	St Elsewhere (UK)
The Drew Carey Show	M*A*S*H	Strong Medicine
Early Edition	Monk	Thirty-something
E R x3	My Wife & Kids	Undressed
Farscape (UK)	The Nanny	The Venture Brothers
Friends	Night and Day	The War at Home
Game On (UK)	Nip/Tuck	The Wayans Brothers
Girlfriends	Northern Exposure	Will & Grace

Unease

Speaking of circumcision in television sitcoms, Glick remarks that, whenever it is discussed, "it is reasonably safe to say" that "the defining motif is uneasiness."[2]

The function of many of the references seems to be (perhaps unconsciously) to resolve that uneasiness, in nervous laughter, dismissal, or anger at anyone who would upset the status quo.

Several themes recur:

1. Infant circumcision is inevitable.
2. Infant circumcision is Jewish.
3. The adult foreskin is disgusting.
4. Circumcision is controversial/important.
5. Circumcision is trivial.
6. Circumcision is safe, quick, and beneficial.
7. Infant circumcision is harder on the parents than the baby.
8. Only *adult* circumcision is painful.

All of these serve to promote infant circumcision.

The paradox of themes 4 and 5 is easily resolved. Circumcision is controversial when anybody suggests *not* doing it, and circumcision is trivial when it's *over*.

I will discuss only the first three themes in detail.

Infant Circumcision Is Inevitable

Genital integrity activist Ari Zighelboim saw this so clearly that he formulated it as a law: "If circumcision of a baby is discussed, baby will be circumcised." Exceptions have occurred only recently.

> In *Ladies' Man*, Jimmy Stiles (Alfred Molina) apologizes to his newborn son that they will be going home to a house full of women (the "sit." of this "com."). A nurse comes in and holds out her arms for the baby. Jimmy unquestioningly hands his son over — and then, as an afterthought, as the nurse is leaving, asks where she is going.
>
> "To get him circumcised."
>
> Without batting an eye, Jimmy says, "And so it begins."[3]

A direct reference like that to the act of circumcising babies in hospital is rare. Routine infant circumcision (RIC) is so invisible that the medical soap, *E.R.*, and the film, *Riding in Cars with Boys*, both showed naked newborn babies who had, it seems, already been circumcised and healed instantly. In *E.R.*, he was brought in after being born in the street; in *Riding in Cars*, he was brought in for his parents to see for the first time, and their surprise was that he was male.

Infant Circumcision Is Jewish

The Jewish proportion of circumcision on television is much higher than in reality. Since Jews comprise 2% of the US population and, assuming 60% of gentile males are circumcised and all Jewish males are circumcised, a maximum 3.3% of circumcision in the US can actually be Jewish.[4] The vast majority of the remainder are neonatal quasi-medical circumcisions and, as Glick[5] points out, many of the Jewish babies are circumcised without ritual in hospitals, and an unknown number are not circumcised (Fig. 1).

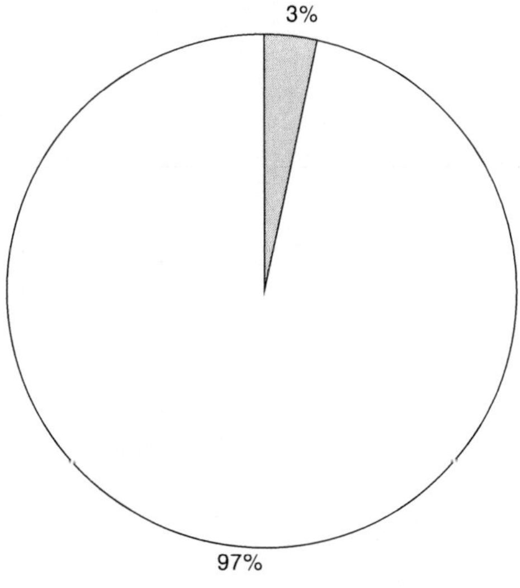

Fig. 1 US circumcision: Jewish fraction

On television, the picture is rather different:

Of 64 references to circumcision in sitcoms, 25 (39%) were Jewish (Fig. 2). When we consider only infant circumcision, the message is even stronger:

Of 41 references to infant circumcision in sitcoms, 19 (46%) were Jewish (Fig. 3).

Presenting the Oscar for special effects in 2006, Ben Stiller pretended to be using green-screen technology to appear as a floating head. His floppy all-over green garment hardly gave proof that Stiller is even male, but host Jon Stewart, said "It's good to have proof that Stiller's Jewish." The near-truism "All Jewish men are circumcised" had become "Only Jewish men are circumcised" in a way whose absurdity would be self-evident in almost any other field. Why it would be "good to have proof" is not clear.

In the context of circumcision and memetics, I said

> The two memes of Jewish circumcision and routine infant circumcision in the US are like the two members of a double star, orbiting each other and influencing each other while keeping their distance — with Jewish circumcision a visible, shining star, routine infant circumcision a black hole, sucking in parents and babies.[6]

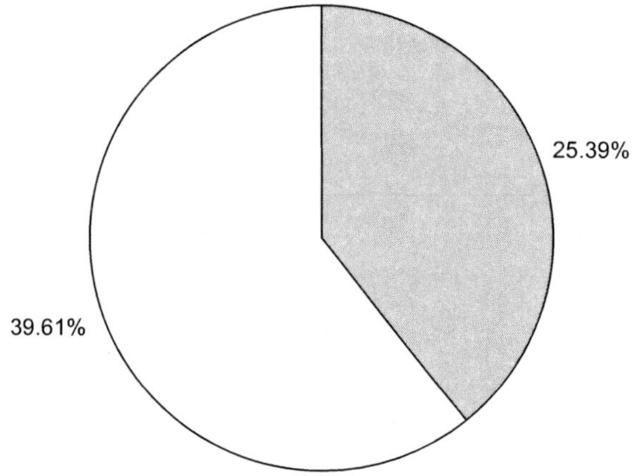

Fig. 2 Circumcision in US TV sitcoms: Jewish fraction (N = 64)

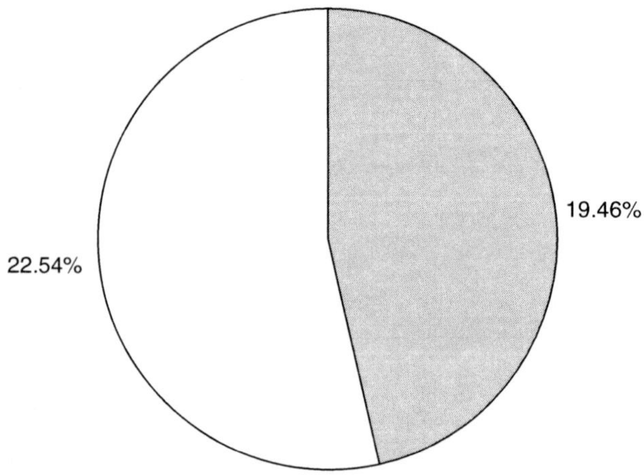

Fig. 3 Infant circumcision in US TV sitcoms: Jewish fraction (N = 41)

This is very clear on television, and the two are virtually never mentioned or discussed together.

The Adult Foreskin is Disgusting

(Glick traces this view back to Mishnah Ned. 3.11, and the exact words of Rabbi Eliezer b. Azariah, which were a Jewish reaction to Christian polemic against circumcision, in the second and third century C.E.[7]) Fourteen out of 64 shows that referred to circumcision (22%) implied that the foreskin is disgusting (Fig. 4).

But, this figure of 22% doesn't do justice to the *depth* of disgust.

- So disgusting that, in one episode[8] of *Off Centre* [*sic*], in which Euan Pearoe (Sean Maguire), a British immigrant to New York, considers being circumcised, other characters make a total of *nine* negative references to his foreskin (Table 2).

And that's one of the rare instances in which circumcision *isn't* inevitable.

- So disgusting that, in *Girlfriends*, when Maya learns a man is intact while folding his washing, she not only screams the obligatory "Ewww!' but flings his *washed* underwear across the room.

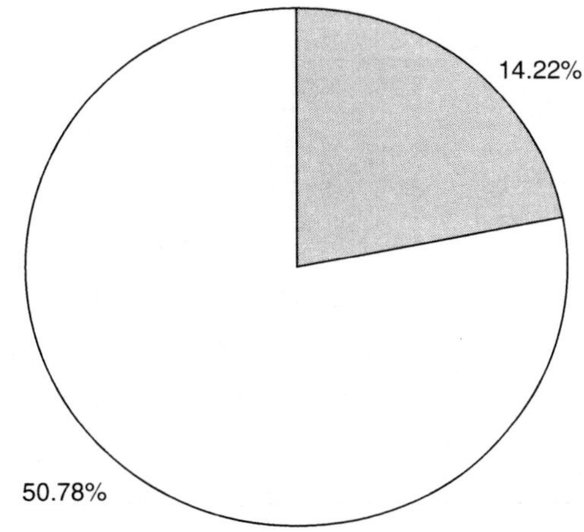

Fig. 4 Proportion of US TV shows mentioning the foreskin that imply it is disgusting (N = 64)

Table 2 Nine negative references to the foreskin in *Off Centre*

1. Chau: (*in terror*) What the hell is that?! On your thingy. You freak!
2. Jay: that penis is a disaster.
3. Dr. Wasserman: Except for the anteater… It's very, very unpleasant.
4. Jordan: Yeah, some chicks are into weird crap.
5. Jordan:(*suddenly yelling*) Euan is uncircumcised! Isn't it gross?
6. Tonya: It's no big deal to me — as long as I don't have to touch it.
7. Mike: What's the matter…they don't like Snuffleupagus?
8. Dr. Wasserman: I can now make a clear diagnosis of your condition: weird wiener-it is.
9. Blonde: (*looks down at his penis*): Whoa!

The *infant* foreskin is seldom referred to. Stories based on infant circumcision seldom go anywhere near what happens to the baby, but focus on how his parents feel, usually on how they overcome their fears and go ahead with it.

Exceptions to all these rules come from outside the US, or are influenced by non-US origins.

The genital integrity movement has been treated at least once, in *Arrested Development* — about a dysfunctional family presided over by a woman who adopts every cause that's going. Unsurprisingly, the real issues are not touched on.

Foreskin restoration is a recent and rare element, always presented as surgery. The non-surgical self-restoration movement is invisible so far.

In More Detail

Sex and the City

In *Sex and the City,* the four upmarket, sexually savvy Manhattan woman friends compare their preference in penises.[9] While one defends intactness (with some disregard for the actual anatomy) –

> I love an uncircumcised dick. It's like a Tootsie-Pop: hard on the outside, with a delicious surprise inside

— the others are scornful:

> I'm sorry, it is not normal.
> Well, actually it is, something like 85 percent of men aren't circumcised.
> Great! Now they're taking over the world!
> Honey, it's a penis, not Godzilla.

— one compares an intact penis to a Shar-Pei.

The issue peaks with the latest man-friend of one of the women having himself circumcised because he has had so many bad reactions from women. The pain of the operation is not stinted (because he's an adult) and she likes the result (though at only a week out, it would actually be a bruised mess), but it's disastrous for their relationship — because circumcision has done his self-esteem so much *good* that he wants to try it out on other women. Though this is New York, there's no reference to Jewish circumcision, and hardly any to operating on babies: to all intents and purposes, a man's foreskin is a rare birth defect.

Seinfeld

By contrast, *Seinfeld*, about four directionless friends, also in Manhattan, treats Jewish ritual circumcision without acknowledging the existence of any other kind.[10] Jerry and Elaine have been chosen to be godparents of a Jewish baby. Remarkably, their neighbor openly opposes infant circumcision, because, he says,

- It's a barbaric ritual.
- A foreskin — referred to only as "it" — makes sex more pleasurable.
- Circumcision "hurts bad."

He even momentarily holds on to the baby and tries to protect it, and he's rewarded for doing so by being made godfather in Seinfeld's place — after the circumcision is over. But the anomalous character is Kramer, who is a kook. In the same episode, he roams the hospital looking for hybrid pig-men, who (or which) he believes are being made there in Frankensteinian experiments, so we may safely discount anything he has to say.

As Glick[11] notes, the circumcision here is Jewish in name only. The *mohel* is a nincompoop who seems to know nothing about the religious role of circumcision. The only reasons given for doing it are:

- An intact penis — never mentioned by name — is ugly, "like a martian"
- Hygiene
- And (the mohel says) it's "an ancient, sacred ceremony, symbolizing the covenant between God and Abraham — or something"

The program makes no direct reference to American routine circumcision, though the gentile characters, George Constanza and Cosmo Kramer, are implicitly circumcised because, when they're asked about intact penises, they refer only to other people.

It exhibits a blindness towards what circumcision entails that would be remarkable if it were not so common. As Glick[11] points out, we hear the onlookers screaming, but not the baby. Jerry's finger is accidentally cut during the ceremony, and he later wails:

> I've never had stitches. I'll be deformed. I can't live with that. It goes against my whole personality. It's not me.

If the scriptwriters planned any ironic reference to Seinfeld's own circumcision in this outburst, it will be too subtle for most people to spot.

Glick[12] remarks, "The inescapable message of this entire performance is that circumcision is perilous. ... Nevertheless, ... Jewish foreskins must be — and will be — removed." I suggest that the purpose of all the nervous laughter around circumcision is to allay people's fears. We might very well call it "gallows humor."

Dharma and Gregg

Free-spirited Dharma and buttoned-down Gregg briefly adopt a boy, and it's mainly Dharma's hippie parents and Gregg's Jewish ones who discuss circumcision, in supposedly comical terms.[13] Eventually they put him through a melange of birth customs, including a Jewish-style circumcision, then reluctantly give him back to his birth mother, at the same time showing her the video of the ceremony. Her only comment: "Who's that fainting?" Once it's done, circumcision is no big deal.

The Simpsons

The most strikingly anti-circumcision message on US TV so far occurs, unsurprisingly, in the iconoclastic animated domestic comedy, *The Simpsons*.[14]

Krusty the Clown finds that he isn't on the Jewish clowns' Walk of Fame because he's never had a Bar Mitzvah, though he bitterly says he was circumcised … "and then some." In his TV show, he regularly runs a cat and mouse cartoon, *The Itchy and Scratchy Show*, that emphasizes gratuitous violence.

> In this episode, called "A Briss before Dying," Scratchy (the cat) is about to circumcise Itchy (the mouse), in a Jewish setting and wearing a *kippah*. Scratchy even utters some words of Hebrew that can be recognized as a commonplace blessing, as before food. But as he holds up the knife, Itchy leaps up and pulls out Scratchy's eyes. Wildly swinging the knife, Scratchy cuts himself to pieces. Itchy puts them through a mincer, forms the mince into a tube, pokes it into a fire and, as if it were glass, blows it into a goblet. He wraps the goblet in a napkin, steps on it and shouts "*Mazel Tov!*" — Congratulations! (a reference to the ceremony at a Jewish wedding).
>
> — all in 35 seconds. Krusty sums up "That's what I believe now."

It's a quite remarkable revenge fantasy, the likes of which we've never seen on flesh-and-blood TV. (It's also a rare episode of *Itchy and Scratchy*, in which the violence is self-defensive.) In light of the actual ratio, and the fact that Itchy and Scratchy have never before shown any sign of being Jewish, it's unexpected that the context is ritual and not surgical. The only reason for that is Krusty's need.

In later episodes of *The Simpsons*,

- Marge Simpson says the dog has to go to the "V. E. T." (presumably to be neutered) and, in the same breath, spells out that son Bart is to be circumcised, implicitly equating the two operations. When Bart says "Huh?" (with the same intonation as the dog), she says she'll explain afterwards. The dog is spared; it's not clear if Bart is.
- Annoying religious neighbor, Ned Flanders, referring to the irreversibility of Bart taking communion as a Catholic, compares it to "the Jews with their (*makes scissoring action with his fingers*) snippety-snip." (Reduplicated speech is one of Flanders' defining characteristics, but here it also suggests hesitation, and perhaps even uneven cutting.)
- Police Chief Wiggum, reading the Constitution of Springfield, mumbles, "… human rights and routine circumcision…" suggesting the topics are linked, but giving no indication how.

The ambivalence of each reference is palpable.

South Park

The other cartoon show that casts any doubt on the wonderfulness of circumcision is *South Park*, with the unequivocal, if not quite accurate, line, "Cuttin' off weewees ain't cool!"

But, after the kids spend most of the episode trying to rescue little Ike from his *Bris*, he treats it so casually — saying "Ouch!" and walking away — they all decide to emulate him. *South Park* and *Off Centre* both joke about it being done with a huge pair of shears.

Off Centre [*sic*]

In *Off Centre,* despite its nine instances of disgusting foreskin, Euan emerges with his penis still intact. (He's talked out of circumcision at literally the last second, by a nurse who chides him for his vanity so, even then, nobody actually defends intactness.) There are many negative references to adult circumcision, including the big shears, and a detailed description of a Gomco circumcision, including separation of the foreskin from the glans — obviously based on the infant operation. The only *intended* reference to neonatal circumcision is also negative but supposedly humorous. If there's a message, it's "Cut them early to save embarrassment later."

Queer as Folk (*US*)

A rare show that treats circumcision at all thoughtfully is the US version of *Queer as Folk*, a sitcom about some gay men, lesbians, and their friends. It's also remarkable in that a biological father, Brian, who is gay, succeeds in saving his son from his *Bris*, again, at literally the last second.[15]

Lindsay: Why does it matter to you if Gus is circumcised?
Brian: It matters that he's been in this world less than a week and already there are people who won't accept him for the way he is. Who would even mutilate him rather than let him be the way he is. The way he was born. Well, I'm not going to let that happen.

Though he even uses the m-word, intactness ("the way he is") is a metaphor for gayness, the main theme of the series.

Ludicrously, the biological mother did not know the *bris* involved circumcision until it was about to happen. It's her lesbian partner, Mel, who is Jewish, and the battle of the sexes takes center-stage.

Lindsay: You know, there are a lot of men who think circumcision is a cruel and barbaric practice.
Mel: I don't care what men think about their dicks! I care that you put Brian before me.

While some remarkably anti-circumcision statements are made, much of the discussion — especially Mel's contribution — ignores any effect circumcision might have on the baby.

In a later episode, a man goes to a gay Jewish function in order find a Jewish husband, and is then rejected at the last second because he's intact. Typically, he offers to convert and be circumcised, but it defies belief that he wouldn't know in advance that it might be an issue.

Married with Children

Married with Children, never the smartest kid on the block, excels itself in an episode[16] in which Al Bundy (Ed O'Neill) is accidentally circumcised in hospital by a doctor who confuses a "circular incision" with a circumcision. Incredibly, nobody suggests suing the doctor or the hospital. (There is a rare but trivializing reference to the pain of routine infant circumcision, in which Al is humiliated by being put into a ward full of crying babies.) Circumcision is trivial, beneficial (to Al's sex life — from being a synonym for male frigidity, he becomes briefly rampant) and a huge joke to his wife. It is painful, but the main disadvantage is the month's enforced celibacy.

E.R.

An episode of the medical soap *E.R.*[17] plays fast and loose with facts to make infant circumcision inevitable. A couple disagree about it at high volume in hospital corridors — the mother accusing the father of wanting his son to look like him (a reason often seriously offered when the father is circumcised, but here, when the father is intact, an *un*acceptable reason). She has the baby cut by another doctor who "likes the look." No-one asks any questions about the medical ethics of this. The father intervenes — too late, and the circumcision has to be completed.

Self-Circumcision

If the idea of circumcision makes viewers uneasy, the idea of doing it to oneself creates a double frisson. At least three shows have dealt with self-circumcision, in each case to please a girlfriend. An underlying message of this plotline is that circumcision is trivial: today a bouquet, tomorrow a foreskin. Men do circumcise themselves in real life, but more commonly for masochistic and/or fetishistic reasons, that is, for sexual pleasure from the operation itself and its contemplation. That is not the kind of theme with which US TV readily engages.

Nip/Tuck

Nip/Tuck, about two cosmetic surgeons in partnership in Miami, revels in the gory details of facial and chest surgery. Several early episodes in the first series deal with the son of the idealistic doctor, convinced that his girlfriend doesn't like his foreskin and unable to persuade the cynical doctor to circumcise him, attempting to do it himself, using instructions he downloads by voice off the Internet. Unusually for this show, we see none of the details. The denouement is that he finds out the girlfriend is lesbian and his foreskin was just an excuse to get rid of him. The operation is completed surgically. If there's a message, it is "Don't try this at home; trust the experts, they know what to do."

E.R. Again

In another episode of *E.R.*, the self-circumcision is completed surgically, only to reveal that the woman didn't really mean it; she too wanted to be rid of the man and thought this would let him down lightly.

House M.D.

In a rare exception to the general rule, Dr. House undoes a young man's attempt at self-circumcision — but with a typically inept euphemism, "putting the Twinkie back in the wrapper."

Game and Talk Shows

Being unscripted, these are an opportunity for common perceptions and misconceptions to be expressed. "Circumcision is Jewish" is one, "Circumcision is minor" another. "The foreskin is disgusting" is less likely to emerge, perhaps because the owner of one may be in a position to answer back but, when he is not, no holds are barred. Thus, Joan Rivers, in her eponymous show, could fall to the ground, look up a dancer's kilt and say, "Please, call a rabbi and have that thing taken care of. I'll pay for it!" (Note the extreme euphemism, "taken care of," and that she refers to a rabbi, not a doctor.)

Outside the US

The attitude on non-US shows is strikingly different. They rarely touch on the subject, but when they do, they are much less biased.

For example, in one episode of the British sitcom *Game On*,[18] ginger-haired agoraphobic Martin Henson (Matthew Cottle) is nervous about his penis in general, its freckles as well as its foreskin. Even that much is unusual in the UK, where circumcision is uncommon. (In several British game and talk shows, men have cheerfully referred to their own foreskins.) But, instead of circumcising himself, he asks his female flatmate to look at it. The euphemisms he uses, "matt" vs "gloss finish" and "Roundhead" vs "Cavalier," imply a degree of equality and familiarity that the US talk of anteaters and Shar-peis does not. After a farcical interruption by their circumcised male flatmate (who predictably misunderstands what is going on), she pronounces his penis "just fine." It is the *circumcised* man who ends up feeling inferior.

Conclusion

While its producers and scriptwriters may not be aware of it, US popular television plays an important role in denigrating the intact penis and reinforcing circumcision as a cultural norm. Over-representation of Jewish ritual circumcision distracts attention from routine infant circumcision and serves to embed circumcision in general as a US cultural practice. Popular misconceptions are reinforced rather than challenged. Factual information is scarce. Since US shows are broadcast worldwide, they may have a role in spreading circumcision — especially infant circumcision — where it is not already prevalent.

Acknowledgment As well as the several references cited, Leonard Glick gave valuable advice and encouragement in adapting this paper from my presentation at the Seattle Symposium.

References

1. Young H. P. The Intactivism Pages: Treatment of Circumcision on TV. Cited 2006 May 20. Available from: URL: http://www.circumstitions.com/TVSitcoms.html
2. Glick, L. B. Marked in Your Flesh: Circumcision from Ancient Judea to Modern America. New York: OUP; 2005, p 241
3. Pilot, Season 1, Episode 1, first broadcast September 20, 1999
4. The population of the US was 300,282,868 on November 26, 2006 (according to http://www.census.gov/population/www/popclockus.html) of whom 6,155,000 (2.05%) were Jewish (according to http://www.jewishvirtuallibrary.org/jsource/US-Israel/usjewpop.html). Assuming the same sex ratio (50:50) for both, and conservatively estimating that 100% of Jewish males and 60% of gentile males are circumcised, that would mean 3,077,500 Jewish males are circumcised and 90,084,860 gentiles, giving a maximum of 3.30% of US circumcision being Jewish.
5. Glick, L. B. Marked in Your Flesh: Circumcision from Ancient Judea to Modern America. New York: OUP; 2005, p 217
6. Young, H. P. Circumcision as a Memeplex. In Hodges, Frederick and Milos, Marilyn, editors. [*Proceedings of the Eighth International Symposium on Human Rights and Modern Society: Advancing Human Dignity and the Legal Right to Bodily Integrity in the 21st Century.*] New York: Kluwer Academic/Plenum Publishers; 2006.
7. Glick, L. B. Marked in Your Flesh: Circumcision from Ancient Judea to Modern America. New York: OUP; 2005, p 47
8. Season 1, Episode 20 "The Unkindest Cut", first broadcast April 21, 2002
9. Series 2, Episode 9, "Old Dog, New Tricks" (or "Old Dogs, New Dicks"), first broadcast August 1, 1999
10. Season 5, Episode 69, "The Bris", first broadcast September 14, 1993
11. Glick, L. B. Marked in Your Flesh: Circumcision from Ancient Judea to Modern America. New York: OUP; 2005, p 267
12. Glick, L. B. Marked in Your Flesh: Circumcision from Ancient Judea to Modern America. New York: OUP; 2005, pp 268–9
13. Season 2, Episode 26, "Turn Turn Turn", first broadcast October 7, 1998
14. Season 15, Episode 6, "Today I am a Clown", first broadcast December 7, 2003
15. Season 1, Episode 3 (or 103), first broadcast December 10, 2000
16. Season 8, Episode 169, "A Little off the Top", first broadcast December 12, 1993
17. Season 4, Episode 72, "Friendly Fire", first broadcast September 9, 1997
18. Series 2, Episode 1 "Roundheads And Cavaliers", first broadcast September 16, 1996

Limbic Imprint

Elena Tonetti-Vladimirova

Abstract Twenty-five years of research in the field of prenatal psychology undoubtedly show a direct correlation between our early experiences in life and the subconscious behavioral and emotional patterns in our adult lives. This mechanism is called "limbic imprint." We now recognize that, however crude our beginning, we do have a choice as adults to reprogram our limbic imprint and transmute suffering and helplessness during the birth process into love and joy of being born on this planet. We can regain our authentic power, clear the pain of our ancestors from our system and set the stage for our children to step into their lives as peaceful empowered guardians of the Earth.

With everything we know about the stages of gestation of the human embryo, it does not bring us any closer to the understanding of that process. Knowing what happens does not really explain why it happens. We can name our body parts and figure out the order of events but we are absolutely mystified in front of that great mystery: What is the force that makes the egg rapidly grow into a human body?

Some things, however, do become increasingly clear — the mechanism of limbic imprinting, for example. Twenty-five years of research in the field of prenatal psychology undoubtedly show a direct correlation between our early experiences in life and the subconscious behavioral and emotional patterns in our adult lives. This mechanism is called "limbic imprint."

To better understand the term "limbic imprint," let us look at the basic structure of our brain. At the tip of the spinal cord, there is a segment called the "reptilian brain," responsible purely for the physiological functions of the body. That is the part of the brain that still remains functional when a person is in a coma; for example, women can even get pregnant while in a coma. Then, there is the cortex, usually referred to as "grey matter," responsible for our mental activity. Finally we have the limbic system of the brain, responsible for our emotions and feelings; this is inevitably where our early experiences are imprinted for the duration of life.

Founder and creator of birthintobeing.com

G. Denniston et al. (eds.), Circumcision and Human Rights,

From conception, through gestation, at birth, and through the first formative years of life, the limbic system thoroughly registers every experience. If we are safe and receive tender, loving care as our first primal experience, our nervous system is "limbicly" imprinted with the undeniable rightness of being — good health, emotional stability, high self-esteem, ability to love, be creative, and responsible.

If our first impressions are anything less than loving, then that "anything" imprints as our "basic settings," acting as a surrogate for the love and nurturing, regardless of how painful it actually was: high levels of stress hormones in the maternal blood stream during pregnancy, inductions or any unnecessary interference with the natural process during labor, lack of immediate contact with the mother after birth, absence of breast milk, immediate severing of the umbilical cord, rough handling in the delivery room, needles, bright lights, loud noises, sexual mutilation — circumcision. All of that sensory overload and excruciating pain becomes instantly wired into our nervous systems as our "comfort zone." Even if later in life our rational mind/cortex will recognize it as a pattern of "abuse," it can seem insurmountable to reprogram these patterns in the areas of life relating to intimacy, trust, and love.

A new baby is an extremely sensitive being — in fact, more sensitive than he or she will ever be during adult life. Babies are not only able to have sensations and feelings, but also to remember them non-cognitively. Our early impressions stay with us for the rest of our life, for better or for worse.

Research done by the pioneers of prenatal psychology, Drs. David Chamberlain, Thomas Verny, and William Emerson, shows that an overwhelming amount of physical conditions and behavioral disorders are the direct result of traumatic gestation and complications during delivery. According to a 1995 study by Dr. William Emerson, 95% of all births in the United States are considered traumatic, 50% rated as "moderate," and 45% as "severe" trauma. Born into excruciating labor pains, numbness and toxicity of anesthesia, we are "limbicly" imprinted for suffering, which strips us of our power and impairs our capacity to experience our true potential. Addictions, poor problem-solving skills, low self-esteem, inability to be compassionate, to be responsible — all of these problems have been linked to birth trauma.

We are familiar with basic settings in our televisions and computers. If a television is set on "maximum blue," then regardless of what is on the screen, it will be blue; or if brightness is on "dim," no matter how bright the image is, the picture on the screen will be dark. It is the same limbic imprint that has been used for thousands of years to train animals: elephants, camels, horses, and circus bears. When a baby elephant is chained to a small stick, it rages for a few days and then stops. When he grows up and has enough strength to pull this stick right out, it doesn't happen. He never tries again. The scientific approach of prenatal psychology that describes that mechanism of limbic imprint exists only in the past 25 years, but it was knowingly and deliberately used for thousands of years among humans by warlords and slave owners for creating soldiers and slaves. When pain and suffering is introduced from the start, it becomes the norm. Bloody and violent aspects of human history were created by people who did not receive the nurturing care they needed as babies. Dr. Stan Grof's statistics that 100% of most violent criminals were unwanted babies speaks for itself. It does not mean that all unwanted babies are bound to become criminals. Most parents who did not want their babies in the beginning manage to rise to the occasion and care for them. But it does mean that those unfortunate ones, whose parents could not meet their needs, have far fewer chances of thriving. The quality of limbic imprint defines our ability to learn and be contented, compassionate human beings, and determines whether living in a body is experienced as a safe and joyful process or a painful and lonesome one.

Most of the masterpieces of human culture reflect the drama of life — jealousy, greed, lust, hate; Shakespeare's plays, for example, are a very beautiful way of speaking about human despair. Only a small fraction of our cultural heritage speaks of love, beauty, and fulfillment. This is because the flow of creativity is also determined by how we feel and experience life, which, in turn, is established by our limbic imprint. It defines our likes and dislikes, what we find attractive and what repels us. In order to give birth to an enlightened masterpiece, whether it would take a form of a baby, a poem, a garden, or simply a rich, fulfilling day that was worth living — one must first heal their own birth trauma. For those of us who were born into a less than ecstatic situation, we need to find a way of healing the trauma that was our driving force from day one. Healthy, loving, self-parenting can neutralize most of the damage. Einstein said: "We cannot solve a problem with the same mindset that created that problem in the first place." We have 250 wars occurring right now around the globe. We have created life-threatening levels of environmental pollution, political systems that do not work, economies that cannot sustain us, and social strategies that ignore our needs. We are clearly, due for some changes. If we truly understand how we created this mess, we have a good chance to rectify it.

Call me naïve, but I truly believe that we can improve the quality of our species in just one generation by allowing our species to enter into this world without being "programmed" to suffering and pain. I envision the new generation coming into the world of safety, compassion, and common sense. If their basic settings are shifted from "anxiety" and "pain" to "love," "safety," and "connectedness," then we Homo sapiens will truly have a chance.

A pregnant woman is nothing short of a miracle worker. She is co-creating a whole new person that did not exist before. If a doctor or nurse approaching a pregnant or laboring woman is able to enter her space with the same frequency as a miracle worker rather than a caseworker, as it usually happens, it would create a very welcoming atmosphere for the baby to make its way into our world.

In a stress-free zone, most of the bodies of the laboring women and their new babies know perfectly well what to do: the built-in mechanism is at work. It is perfectly designed for producing a new human being. All of the necessary chemical and physical adjustments inside the woman and baby's bodies will happen with appropriate timing — with one condition; all the obstacles need to be removed. Birth cannot be taken out of the context of life. It does not start with labor. It starts with conception. The quality of gestation will greatly define the quality of delivery. Most of the complications can be prevented during pregnancy through preparation for conscious birth, which includes emotional and physical healing, self-empowerment and a great deal of learning.

I made a documentary, *Birth As We Know It,* about this very subject. There are more than three and a half hours of material included in the DVD, which delivers a powerful transmission of what it really takes to give birth consciously and gracefully. The film features 11 stunningly beautiful natural births and interviews on topics rarely discussed, such as the sexuality of childbirth, harmful consequences of cesarean section, and circumcision, yet primarily it reveals fascinating insights into the preparation necessary for natural birth — most important of which is to release the unconsciously held trauma from both parents' own births.

We can make a conscious effort to heal our own birth trauma and embrace the opportunity to create a masterpiece of our life. We can recognize that, however crude our beginning was, we do have a choice as adults to reprogram our limbic imprint and transmute suffering and helplessness during the birth process into love and joy of being born on this planet. We can regain our authentic power, clear the pain of our ancestors from our system, and set the stage for our children to step into their lives as peaceful empowered guardians of the Earth.

Two Poems

If we hope to create
a non-violent world
where respect and kindness
replace fear and hatred,
we must begin
with how we treat each other
at the beginning of life.

For that is where
our deepest patterns are set.
From these roots
grow fear and alienation
– or love and trust.

~***~***~***~

When we have learned
to love, respect, and protect
the integrity of every child's
body, mind, and spirit...
Then we will have reached

a new era in human consciousness.
And our children will thank us
by respecting themselves,
each other, and mother earth.
That is where we are heading.

Suzanne Arms

G. Denniston et al. (eds.), *Circumcision and Human Rights*,

Appendix A

In The Circuit Court of Cook County, Illinois County Department, Domestic Relations Division

In re the former marriage of:	)	
	)	
Adriana Schmidt,	)	
Petitioner,	)	
	)	
and	)	No. 00 D 18272 cons.
Marian Niznik,	)	00 D3 31904
Respondent.		

Respondent's Closing Argument

As requested by the Court at the close of the evidence, Respondent/Movant Marian Niznik herewith submits the following as his closing argument:

This is a simple case. Ms. Rovin, formerly Ms. Schmidt, was required by the divorce decree to confer with Mr. Niznik about any non-emergency healthcare services for the Child. (Parenting Agreement, Paragraph B, attached to the Judgment for Dissolution of Marriage, Exhibit "B" to Respondent's Emergency Verified Petition herein). She failed to do so, both in 2005 and in 2006. Instead she secretly scheduled an unnecessary circumcision of the Child, even though the child had never been properly diagnosed and treated for alleged "balanitis," or inflammation of the glans penis. Mr. Niznik discovered this plan, discussed the facts about circumcision with his son (who upon being informed of the truth decided he did not want to be circumcised), and forbade the circumcision. He then brought this action to obtain the Court's aid in prohibiting the circumcision and to have Ms. Rovin held in contempt for failing to confer with him. The undisputed facts support his requests, particularly since the undisputed facts at the evidentiary hearing were (1) that the Child at present has an entirely normal, disease free penis, (2) that circumcision is physically damaging, (3) that circumcision at this age may be psychologically damaging, (4) that circumcision at the Child's age carries with it the risks inherent in the use of general anesthesia, including death, (5) that circumcision carries with it the risk of surgical mishap, bleeding, and infection, with possible disastrous consequences, (6) that balanitis is almost always 100% curable by the application of betamethasone cream, which has never been prescribed for the Child, and (7) that balanitis can occur in a circumcised male.

First, it is indisputable that Ms. Rovin intentionally did not tell Mr. Niznik about the proposed circumcision before she scheduled it. Mr. Niznik only found out about the scheduled surgery a few days before it was to occur in February 2006, when his son, during a scheduled visitation period, told him he was to have an operation on his penis that would leave him feeling "freaky" for a couple of days. Mr. Niznik then confronted Ms. Rovin with his discovery, but she would not relent.[1] While Ms. Rovin now insists that the Child suffered three bouts of "balanitis" during 2005, she did not tell Mr. Niznik about them, nor did his son ever mention them to him. Even if they occurred, they were unlikely to have been as painful as Ms. Rovin related at trial because (1) in order to have been so painful the foreskin would have had to have been seriously eroded and raw, a condition no one has described,[2] and (2) Dr. Goldstein did not believe they were significant enough to require an office visit, nor did he believe that they were significant enough to note in the Child's medical record. (Defendant's Exhibit 55 "D-55"). In all events, each episode cleared after the application of Neosporin, recommended by Dr. Goldstein, despite the fact that according to Dr. Gibbons Neosporin is a known irritant to some patients and is not indicated for use on the glans penis for this reason.

Likewise the Child's January 2006 bout of balanitis cleared despite the application of prescribed Bactroban, which likewise is not indicated for use on the glans penis as it does not attack any of the microbes that attack the glans. Indeed, Dr. Kaplan, who did not find the Child to have balanitis, apparently recommended circumcision based

---

[1] Ms. Robin's protestations at trial that she tried to tell Mr. Niznik about the circumcision in the very telephone call where he brought the subject up should be rejected as a self-serving attempt to avoid the legal consequences of her contemptuous refusal to follow the Court's order incorporating the parties' agreement. Likewise her contention that only a "temporary" decision had been made to circumcise is ludicrous given the fact that Mr. Niznik had to threaten a lawsuit to keep the doctor from performing the circumcision and had to then bring this action.

[2] The truth of this may be seen from noting that the foreskin has both an epidermis and dermis layer on the outside, see D-5, D-13, D-14, D-16, D-17, D-18 and D-19. If worn forward in the normal anatomical position the only way a boy or man could not stand to have it touch underwear would be if the outer layers were seriously eroded or had suppurating lesions. These were never described by anyone. It is far more likely that Ms. Rovin tried to have the Child wear his foreskin in as far a retracted position as possible with his glans partially exposed. Given the un-keratinized nature of the mucosal surface of the glans of an uncircumcised penis, a child with his foreskin pulled back, even partially, would feel pain as the glans rubbed against underwear. This may partially explain the Boy's alleged complaints of pain, although since the Boy never complained to his dad of any pain or discomfort, despite seeing him every Tuesday for visitation, it is more likely that Ms. Rovin has exaggerated the alleged symptoms. Of course, the other alternative is that the Child never had balanitis in 2005. This is possible since Ms. Rovin admitted calling Mr. Niznik in January 2006 to tell him the Child had an ointment to put on his penis. She would not have made this call if the Child had been using Neosporin on his penis as often as she claimed he did in 2005. Further Dr. Goldstein's records do not reflect any discussion of balanitis in 2005, nor do they reflect the recommendation of the application of Neosporin (D-55).

only on related past history and on the presence of a foreskin that experts for both parties agree is in a normal state of health and development.[3]

Second, it is indisputable, as testified to by Drs. Van Howe and Gibbons,[4] that circumcision is damaging physically because it removes the foreskin, which occurs naturally in all males and by its very nature protects the glans penis from injury. Further it is now known that the foreskin is important for full sexual satisfaction of both male and female partners because it contains most of the fine touch nerve receptors in the penis and during intercourse sits behind the corona glandis and stimulates both partners (see D-20, D-23). Further circumcision at the age of eight to nine, the age of the Child, carries with it a risk of death or brain damage from necessary general anesthesia, a risk of surgical damage, and a risk of bleeding, infection, and permanent loss of the penis (see D-31, D-32, D-33, D-34, D-35, D-36, D-37, D-38, D-39, D-40, D-41, D-42, D-43, D-44, D-45, D-46). It is almost never medically necessary to circumcise a boy of the Child's age, even one with recurrent balanitis. Balanitis, which in a young boy is almost always caused by physical irritants, particularly those high in alkalinity (pH) such as concentrated urine, soap, and swimming pool water, and which rarely is caused by a virus or bacterium at this age, can almost always be cured by the application of betamethasone cream.[5] The only medical indication for circumcision of an 8 to 9 year old boy is BXO, which the Child does not have. Finally circumcision at this age carries with it the risk of serious psychological damage.[6]

---

[3] This includes Dr. Van Howe, who found the child to have a normal, non-diseased foreskin in March 2006, and Dr. Hatch, Ms. Rovin's testifying urologist, who found the Child to have a normal, non-diseased foreskin in May 2006. Dr. Kaplan, who did not testify, apparently based his recommendation for circumcision on the existence of normal connections between the Child's foreskin and his glans, a condition that will change as the Child matures and the connections naturally dissolve. Such connections are normal at this age according to Drs. Van Howe, Gibbons, and Hatch and do not require circumcision (Dr. Van Howe noted that many otherwise fine physicians may not be current on foreskin anatomy and function). Dr. Goldstein apparently was entirely unfamiliar with the last 20 years of medical literature in regard to the normal development of the penis from birth to adolescence. Therefore his opinion in regard to the circumcision of the Child is suspect at best.

[4] Neither Dr. Goldstein, nor Dr. Hatch denied the function of the foreskin, its innervation, or its sexual function. Neither denied the risks of circumcision, although they may have downplayed them. Neither stated that circumcision was *medically necessary* for this Child. Neither appeared very familiar with the medical literature on psychological damage caused by genital surgery at the Child's age, a fact testified to by both Dr. Gibbons and Dr. Van Howe.

[5] Only Dr. Gibbons has had experience using betamethasone to treat balanitis and he has found it to be almost 100% effective. Dr. Hatch testified he has never used it to treat balanitis.

[6] The possibility of psychological damage here seems particularly acute since the Child has written his mother that he does not want to be circumcised. While the mother said the Child was hysterical when he came home from the visit where his father discussed exactly what circumcision consists of with him, the Child was not hysterical when the father had the discussion with him. It seems likely that the hysteria was caused by the Child's realization that his mother would not cancel the circumcision after he expressed his desire not to be circumcised. Such a reaction is to be expected from anyone, and is only a precursor of the anger, resentment, and hatred the Child will more than likely feel toward his mother and step-father if the circumcision now takes place.

Indeed, two points are noteworthy here. The first is the fact that each time the Child apparently contracted balanitis he did so after swimming in a swimming pool. (His step-father admitted that in May 2005 they went to heavily chlorinated swimming pools.) Since swimming pools must constantly have their water chemistry balanced, it seems obvious that the high pH of the water may have been the cause of the irritation of the Child's penis in each case. Avoiding high pH swimming pools would effectively eliminate the cause of the Child's balanitis without subjecting him to painful, damaging surgery. However, no one, including Drs. Goldstein, Kaplan, the Florida doctor, or Dr. Hatch, ever undertook to medically diagnose the cause of the Child's balanitis before recommending circumcision.[7] Therefore the cause is unknown. Second, what is known, however, is that each episode cleared within one (1) week without surgical intervention. Whether or not the balanitis was caused by swimming pool water, the use of soap under the foreskin, temporarily concentrated urine or even if it was compounded by the use of Neosporin, which is known to irritate certain children, no surgery was necessary to obtain complete relief. Since February the Child has not suffered another bout of balanitis. He presently has a healthy, normal penis, a fact no one disputes. There is no reason for surgery. Indeed, since even circumcised men can get balanitis, there is no guarantee that circumcision would prevent a re-occurrence of the Child's prior problems.

This is probably why Dr. Hatch was forced to admit under cross-examination that it is not essential to the present health of the child that he be circumcised! Given this admission, this case should be at an end and circumcision should be enjoined.

Third, the Court should note that in such a case it is medically unethical to recommend that the Child be circumcised. Indeed, it is a violation of medical ethics to circumcise a child of this age in these circumstances when he has not given his assent, but rather has withheld it.[8] As both Drs. Van Howe and Gibbons opined, it would be beneath the applicable standard of care to circumcise a child like this one who presently has a healthy uncircumcised penis and who does not assent to the procedure. While Drs. Hatch and Goldstein might feel it is "better" to remove the foreskin,[9] the Court must protect the Child from permanently damaging surgery that changes the appearance and functioning of his body when that surgery is not essential to the *present* health and well-being of the Child and he has not assented thereto.

---

[7] This was a breach of medical ethics, since surgery that removes a normal body part should only be undertaken after medical treatment has failed or after a diagnosis that makes it plain that medical treatment would be ineffective.

[8] See AAP Bioethics Committee Statement (D-30).

[9] Each seems to have a pro-circumcision prejudice. Dr. Goldstein testified he recommends routine neonatal circumcision and disagrees with the American Academy of Pediatrics statement on neonatal circumcision although without any stated basis. Dr. Goldstein admitted that he is ignorant of the sexual functions of the foreskin. Dr. Hatch testified he performs 250 circumcisions a year, about 20 of those on 5 to 10 year olds. This is in stark contrast to Dr. Gibbons, a highly qualified pediatric urologist, who only circumcises if there is a grave risk to the health of the child and who has only done perhaps 40 or so circumcisions in his career. Dr. Hatch does not have knowledge or opinions about the long term psychological effect of circumcision of a child of this age.

Before a Court may authorize a surgical procedure on a child it must be shown that the same is (1) medically *necessary* and *urgent* (2) that there are no reasonable alternatives, and (3) that the contingencies or risks are minimal. Cf. In Re Richardson, 284 So.2d 185 (La. 1973). This is certainly in keeping with the "best interests of the child." The "best interests" of the child are never served by having unnecessary, non-emergency surgery performed upon him when there are reasonable alternatives to the procedure and where the risks are more than minimal if surgery is performed.[10] Further, medical ethics prohibits circumcision in a case like this one. Margaret Somerville, a well-known medical ethicist and lawyer at McGill University in Montreal, Canada, has written the following in regard to routine, non-religious infant circumcision, but the analysis is equally applicable to the circumcision of a healthy 8–9 year old uncircumcised boy:

> A common error made by those who want to justify infant male circumcision on the basis of medical benefits is that they believe that as long as some such benefits are present, circumcision can be justified as therapeutic, in the sense of preventive health care. This is not correct. A medical-benefits or "therapeutic" justification requires that overall the medical benefits should outweigh the risks and harms of the procedure required to obtain them, that this procedure is the only reasonable way to obtain these benefits, and that these benefits are necessary to the well-being of the child. None of these conditions is fulfilled for routine infant male circumcision. If we view the child's foreskin as having a valid function, we are no more justified in amputating it than any other part of the child's body unless the operation is medically required treatment and the least harmful way to provide that treatment.

Somerville, M., *The Ethical Canary: Science, Society and the Human Spirit*, pp. 204–205 (Viking 2000).[11] Here since there is no medical emergency, since at present (and at least since February of this year) the Child has a normal, non-diseased, non-inflamed uncircumcised penis, since any purported benefits (i.e. less future bouts of balanitis) not only have not been proven but are mere conjecture,[12] since the risks of injury and death are not *de minimis* but rather are very real, since the risk of psychological harm at this age is very high, particularly since the Child does not want to be circumcised, since the physical losses and changes in sexual function and sensation caused by circumcision are indisputable, since each bout of balanitis has gone away *within one week* without surgery, and since the application of betamethasone cream is nearly 100% effective if treating balanitis without surgery, it is not in the best interests of this Child that he be circumcised. Therefore, the Court must protect him from the attempt to do so.

---

[10] While Drs. Hatch and Goldstein tried to downplay the risks of circumcision, the fact that Dr. Gibbons sees so many circumcised boys needing corrective surgery more than proves that circumcision is not without substantial risk of injury.

[11] Further detailed discussion in this regard may be found in Respondent Marian Niznik's Pre-Trial Memorandum of Law and Points and Authorities in Support of his Emergency Petition for Temporary Restraining Order and/or Preliminary Injunction Prohibiting the Circumcision of N.N. heretofore filed in this action, which is incorporated herein by reference.

[12] Particularly since it is undisputed that circumcised men get balanitis upon occasion (indeed as Dr. Van Howe noted circumcised infants under the age of three get balanitis more often that uncircumcised infants of the same age) and that circumcision is no guarantee that balanitis will not recur.

In sum, the Court must enter an order enjoining Ms. Rovin from consenting to the Child's circumcision without further order of the Court after an evidentiary hearing has proven the necessity of the circumcision and enjoining the whole world from consenting to the Child's circumcision or from performing a circumcision on the Child until further order of the Court. (see 750 ILCS 5/602 and 5/608). Further, the Court must find Ms. Rovin in intentional contempt of the final judgment and decree and must award Mr. Niznik's attorney's their fees and expenses for having to bring this action.

This _____ day of July, 2006.

David J. Llewellyn (*Pro hac vice*)

Johnson & Ward
2100 The Equitable Building
100 Peachtree Street, NW
Atlanta, GA 30303-1912
Tel: 404-524-5626
Fax: 404-524-1769

John M. D'Arco

Michael G. DiDomenico

Lake, Toback & D'Arco
Attorney No. 91154
33 North Dearborn Street, Suite 900
Chicago, Illinois 60602
Telephone No. (312) 726-7111
Facsimile No. (312) 726-2385

Attorneys for Respondent Marian Niznik

Appendix B

Resources

Organizations

Association Contre la Mutilation des Enfants (A.M.E.). Didier Diers and Xavier Valle, Boite Postale 220, 92108 Boulogne Cedex, France. http://pages.pratique.fr/~ame1/

Attorneys for the Rights of the Child. J. Steven Svoboda, JD, 2961 Ashby Avenue, Berkeley, CA 94705 USA. Tel: 510-595-5550. www.arclaw.org

Circumcision Information Australia. www.circinfo.org/

Circumcision Resource Center. Ronald Goldman, PhD. PO Box 232, Boston, Massachusetts, 02133 USA. Tel: 617-523-0088. www.circumcision.org/

Doctors Opposing Circumcision (D.O.C.). George Denniston, MD, MPH, President; John Geisheker, JD, Executive Director.
www.doctorsopposingcircumcision.org
The Prepuce by Steve Scott
www.doctorsopposingcircumcision.org/video/prepuce.html

Equality Now. Jessica Neuwirth, President. PO Box 20646, Columbus Circle Station, New York, NY 10023. Tel: 212-586-0906. Fax: 212-586-1611. www.forward.org

Foundation for Women's Health Research and Development (FORWARD). Sarah Fisher, 765–767 Harrow Road, London NW10 5NY. Tel: 020-8960-4000. Fax: 020-8960-4014.

Inter-African Committee. Berhane Ros-Work, President. 147 rue de Lausanne, CH-1202 Geneva, Switzerland. Tel: 22-731-2420. Fax: 22-738-1823.

International Centre for Reproductive Health. Els Leye, FGM Project Coordinator. Ghent University, De Pintelaan 185 P3, 9000 Ghent, Belgium. Tel: + 32-9 240.35.64. Fax: + 32-9 240.38.67.

International Coalition for Genital Integrity. Dan Bollinger. Tel: 765-427-7012. www.icgi.org

Israeli Association Against Genital Mutilation. Avshalom Zoossmann-Diskin. PO Box 56178. Tel-Aviv 61561 Israel. www.britmilah.org

London Black Women's Health Action Project. Shamis Dirir. Cornwall Avenue Community Centre, First Floor, 1 Cornwall Avenue. London E2 0HW United Kingdom. Tel: 181-980-3503. Fax: 181-980-6314.

Medical Ethics Network. John Sawkey, PO Box 578, Yorkton, Saskatchewan, S3N 2W7. Tel: 306-744-2436. http://med-fraud.org

National Organization of Circumcision Information Resource Centers (NOCIRC) (International Headquarters) Marilyn Fayre Milos, RN, Executive Director. PO Box 2512, San Anselmo, CA 94979-2512. USA. Tel: 415-488-9883. Fax: 415-488-9660. www.nocirc.org/

National Organization to Halt the Abuse and Routine Mutilation of Males (NOHARMM). Tim Hammond. www.noharmm.org/

National Organization of Restoring Men(NORM) International Headquarters. R. Wayne Griffiths, MS, Med, 3505 Northwood Drive, Suite 209, Concord, CA 94520-4506 USA. Tel: 510-827-4066. Fax: 510-827-4119. www.norm.org/

NORM-UK. John P. Warren, MB. Chairman. PO Box 71. Stone, Staffordshire, ST15 0SF, United Kingdom. Tel/Fax: 01785-814-044. www.norm-uk.co.uk/

Nurses for the Rights of the Child. Mary Conant, RN, Betty Katz Sperlich, RN, Mary-Rose Booker, RN. www.cirp.org/nrc/

Rainb♀. Nahid Toubia, MD. 915 Broadway, Suite 1109, New York, NY, 10010-7108 USA. Tel: 212-477-3318. Fax: 212-477-4154

Terres des Femmes. Petra Schnull, Gritt Richter, Claudia Piccolantonio. Kreuzbergring 10, D-37075 Göttingen, Germany.

WorldWide Web Sites

Alliance for Transforming the Lives of Children
www.atlc.org/
Association Contre la Mutilation des Enfants (French)
http://pages.pratique.fr/~ame1/
Attorneys for the Rights of the Child
www.arclaw.org/
Birth Psychology
www.birthpsychology.com/birthscene/circ.html
BoysToo.com (Official Website of NOCIRC of North Dakota)
www.boystoo.com
Circumcision and HIV
www.circumcisionandHIV.com/
Circumcision Information and Resource Pages
www.cirp.org/
Circumcision Information Resource Center (Montreal, Canada)
www.infocirc.org/index-e.htm
Circumcision Resource Center (Boston, Massachusetts)
www.circumcision.org/

D.O.C. (Doctors Opposing Circumcision)
www.doctorsopposingcircumcision.org
Female Genital Mutilation Research Home Page
www.fgmnetwork.org/
In Memory of the Sexually Mutilated Child (John A. Erickson)
www.datasync.com/SexuallyMutilatedChild/
The Intactivism Pages
www.circumstitions.com
International Coalition for Genital Integrity
www.icgi.org/
Intersex Society of North America Home Page
www.isna.org/, www.dsdguidelines.org
Jews Against Circumcision
www.JewsAgainstCircumcision.org
National Organization of Circumcision Information Resource Centers
www.nocirc.org/
National Organization to Halt the Abuse and Routine Mutilation of Males
www.noharmm.org/
National Organization of Restoring Men (NORM)
www.norm.org/
NORM-UK (Great Britain)
www.norm-uk.org/
Nurses for the Rights of the Child
www.cirp.org/nrc/
Students for Genital Integrity
www.studentsforgenitalintegrity.org/

Books

Sami A. Aldeeb Abu-Sahlieh. *Male and Female Circumcision Among Jews, Christians and Muslims: Religious Debate*. Beirut, Lebanon: Riad El-Rayyes Books. 2000.

Sami A. Aldeeb Abu-Sahlieh. *Circoncision Masculine — Circonsion Femine: Debat Religieux, Medical, Social et Juridique*. Paris: L'Harmattan. 2001.

Jim Bigelow, PhD. *The Joy of Uncircumcising!* 2nd Edition. Aptos, CA: Hourglass. 1995. [ISBN 0-934061-22-x]

Elizabeth Heger Boyle, *Female Genital Cutting: Cultural Conflict in the Global Community*. Baltimore. The Johns Hopkins University Press. 2002.

Robert Darby. *A Surgical Temptation: The Demonization of the Foreskin & the Rise of Circumcision in Britain*. Chicago: The University of Chicago Press. 2005.

George C. Denniston and Marilyn Fayre Milos, eds. *Sexual Mutilations: A Human Tragedy*. New York and London: Plenum Publishing Corporation. 1997.

George C. Denniston, Frederick Mansfield Hodges, and Marilyn Fayre Milos, eds. *Male and Female Circumcision: Medical, Legal, and Ethical Considerations in Pediatric Practice*. New York: Kluwer Academic/Plenum Publishers.1999.

George C. Denniston, Frederick Mansfield Hodges, and Marilyn Fayre Milos, eds. *Understanding Circumcision: A Multi-Disciplinary Approach to a Multi-Dimensional Problem*. New York: Kluwer Academic/Plenum Publishers. 2001.

George C. Denniston, Frederick Mansfield Hodges, and Marilyn Fayre Milos, eds. *Flesh and Blood: Perspectives on the Problem of Circumcision in Contemporary Society*. New York: Kluwer Academic/Plenum Publishers. 2004.

George C. Denniston, Pia Grassivaro Gallo, Frederick M. Hodges, Marilyn Fayre Milos, and Franco Viviani, *Bodily Integrity and the Politics of Circumcision: Culture, Controversy, and Change*, Springer. 2006.

Efua Dorkenoo. *Cutting The Rose: Female Genital Mutilation: The Practice and its Prevention*. London: Paul & Co Pub Consortium. 1996.

Paul M. Fleiss and Frederick M. Hodges. *What Your Doctor May Not Tell You About Circumcision*. New York: Warner Books. 2001.

Leonard B. Glick. *Marked in Your Flesh: Circumcision from Ancient Judea to Modern America*. New York: Oxford University Press. 2005.

Ronald Goldman, PhD. *Circumcision: The Hidden Trauma*. Boston: Vanguard. 1996.

Ronald Goldman, PhD. *Questioning Circumcision: A Jewish Perspective*. Boston: Vanguard. 1997.

David L. Gollaher, PhD. *Circumcision: A History of the World's Most Controversial Surgery*. New York: Basic Books. 2000.

Pia Grassivaro Gallo and Franco Viviani, eds. *Female Genital Mutilation: A Public Health Issue Also in Italy*. Proceedings of the 1994 International Symposium on Female Genital Mutilation, May 3rd, 1994, Padua, Italy. Padua: UNIPRESS. 1995.

Ellen Gruenbaum. *The Female Circumcision Controversy*. Philadelphia: University of Pennsylvania Press. 2000.

Fadumo Korn . *Born in the Big Rains*, Feminist Press at the City University of New York. 2006.

Hanny Lightfoot-Klein. *Prisoners of Ritual: An Odyssey Into Female Genital Circumcision in Africa*. New York: Harrington Park Press, 1989.

Hanny Lightfoot-Klein. *Children's Genitals Under the Knife: social imperatives, secrecy and shame*. BookSurge Publishing. 2007.

Peggy O'Mara, ed. *Circumcision: The Rest of the Story*. Santa Fe, NM: Mothering.1993.

Caroline Omoifo. *Saving Bekyah: Confronting Female Circumcision, Sexuality, and Womanhood*. Sun Rose Publishers. 2007.

Anika Rahman and Nahid Toubia. *Female Genital Mutilation: A Guide to Laws and Policies Worldwide*. London: Zed Books, 2000.

Thomas J. Ritter, MD, and George C. Denniston, MD. *Doctors Re-examine Circumcision*. 3rd Edition. Aptos, CA: Hourglass, 2002.

Rosemarie Skaine. *Female Genital Mutilation: Legal, Cultural and Medical Issues*. North Carolina: McFarland & Company, Inc., Publishers, 2005.

Margaret Somerville. *The Ethical Canary: Science, Society and the Human Spirit*. Toronto: Penguin Books, 2000.

Edward Wallerstein. *Circumcision: An American Health Fallacy*. New York: Springer, 1980.

Kayla Weiner, PhD, and Arinna Moon, MA., eds. *Jewish Women Speak Out: Expanding the Boundaries of Psychology*. Seattle, WA: Canopy Press, 1995.

Videotapes/Films

Circumcision? Intact Facts. 18-min. VHS. $44.05. Injoy Productions, 1435 Yarmouth, Suite 102-B, Boulder, CO 80304. Tel: 800-326-2082.

The 8th Day. 53 min. VHS. $30ppd in US, $50ppd outside US. Keren Markuze, POB 361425, Los Angeles, CA 90036. Tel: 323-936-6802.

Facing Circumcision: Eight Physicians Tell Their Stories and Reveal the Ethical Dilemmas of Physicians who Circumcise Newborns. Nurses for the Rights of the Child. 20 minutes. VHS. 1998. Nurses for the Rights of the Child. 369 Montezume #354, Santa Fe, New Mexico, 87501. Tel: 505-989-7377. www. cirp.org/nrc/.

Fire Eyes. Soraya Mire. 60 minutes. Sale: 16mm $2,000, video $445, rental: 16mm $300, video $85. Filmakers Library, Inc., 124 East 40th Street, New York, Y 10016. Tel: 212-808-4980.

It's a Boy! Victor Schonfeld. 41 minutes. VHS. $295 institutions, $195 individuals, $65 rental. Filmmakers Library, 124 East 40th Street, New York, NY 10016. Tel: 212:808-4980. Fax: 212-808-4983.

NOCIRC PSA and Educational DVD. 20 minutes. $5. NOCIRC, POB 2512, San Anselmo, CA 94979-2512.

The Nurses of St. Vincent: Saying "No" To Circumcision. Barry Ellsworth. Includes *Tahara*. Sara Rashad. 18 min. NTSC VHS. $30 ($200 for institutions, schools, libraries). Order at www.tahara-film.com.

Whose Body, Whose Rights? Lawrence Dillon and Tim Hammond. 1996. 56 minutes. VHS. Home Sales: Video-Finders, 1-800-343-4727. Educational Sales: $195, Rental $70, Catalogue no. 38342, University of California Extension, Center for Media and Independent Learning, 2000 Center Street, Fourth Floor, Berkeley, CA 94704. Tel: 510-642-0460.

Cut: Slicing Through the Myths of Circumcision. Eliyahu Ungar-Sargon. DVD. $24.95 plus $4.95 S&H. www.CutTheFilm.com.

Newsletters

Attorneys for the Rights of the Child Newsletter. Albert Fields, Editor. ARC, 2961 Ashby Avenue, Berkeley, CA 94705, USA.

Awaken. Faiza Jama Mohamed, Editor. Equality Now, 250 W 57th St #1527, New York, NY 10107, USA.

NOCIRC Annual Newsletter. Marilyn Fayre Milos, RN, Editor. NOCIRC, POB 2512, San Anselmo, CA 94979-2512, USA.

NOCIRC of Michigan Informant. Norm Cohen, Editor. POB 333, Birmingham, MI 48012, USA.

NORM NEWS. David Smith, Editor. NORM-UK, POB 71, Stone, Staffordshire, ST15 0SF, England.

Index